Insights

2

Biology
and ethics

Reflections inspired by a Unesco symposium
by Bruno Ribes

The opinions expressed in this work
are those of the author and are not
necessarily those of Unesco.

Published in 1978 by the United Nations
Educational, Scientific and Cultural Organization
7 Place de Fontenoy, 75700 Paris
Printed by Sydenhams Printers, Poole, Dorset

ISBN 92–3–101568–0
French edition: 92–3–201568–4
Spanish edition: 92–3–301568–8

Preface

In order to constitute itself as a science, biology had to work out its own specific concept of life, defending the independence of its methods as distinct from those of physics and chemistry. The biologist sought to enter into what nature had already, as it were, set up as an experiment in living matter, rather than to intervene, abruptly or otherwise, in the cause of its development. He was not in any case equipped with the means for intervening in living matter without disturbing it, and ran the risk of destroying the subject of his study by modifying its natural state. Pathological exploration was the biological equivalent of experiments in physics or chemistry.

At the present time, in order to proceed with its analysis of the structure of living beings, biology has had to have recourse to the closest co-operation with physics and chemistry, at the degree of refinement they have currently reached: the result has been the appearance of molecular biology.

Molecular biology probably makes no claim to defining the nature of living beings, but it enlightens us on their constituent parts. It investigates in depth the manner in which they are organized and reveals their minutest details, at the level of the individual cell, or, better still, of the structure of the molecules which characterize the cell.

But this is a level to which one can only have access by employing all the methods and techniques made available by biochemistry, modern physics, genetics, physiology, etc., and by having recourse to the most sophisticated observation techniques. Molecular biology is thus situated at the intersection of a great number of disciplines, and is essentially of an interdisciplinary character. It can even be said that, in describing the correlations between the internal chemical structure of the molecular nucleus (the chromosome) and the global phenotype of an organism, genetics belies the traditional separation of the sciences, and combines together in a new discipline principles and laws which lie between different levels; in so doing, it discovers uniformity in two kinds of phenomena, and makes the central point of its programme the problem of the relation between heredity and evolution. Thus the hiatus is filled which in classical genetics separated the gene, the ultimate term in the analysis and an element in heredity, from such characteristics as, for example, the form of the seeds or the position of the flowers.

For the findings of a precise chemistry and of a rigorous physics frequent twists and turns, each corresponding to a new technique or a new piece of knowledge, to be gradually built up into a thematic unity involves a lengthy process with a constant shuttling between the cell in its intact state and the cell broken down for the purposes of analysis, a movement which makes it possible to verify whether the artificial, experimental conditions correspond to the natural conditions of experience.

The result is a knowledge of genetic material which confers the ability to intervene in life. The key to this ability is of course the experimental nature of biology: there is in fact nothing in principle to prevent the knowledge of genetic material gained through experiment from being used for the purpose of acting on it, or from revealing its transformational capacity. This is one of the numerous problems posed by modern biology, and lies at the heart of the questions discussed in the following pages, the author of which focuses attention on the misgivings raised among scientists themselves by the new powers derived from new forms of knowledge.

The present publication marks the convergence of three different lines of approach.

In the first place, there is the development of molecular biology itself, the heir to the experimental biology founded by Virchow, Darwin, Claude Bernard, and also Mendel, Pasteur and Berthelot, followed by de Vries, Morgan, Weismann, Miescher, etc., to quote only the best known. This succession of names gives no indication of the complexity of the paths followed, with their inherent pitfalls, in the search to elucidate a problem which was exceptionally complicated both on account of its subject matter and the resources required for its study. To take only one example, Pasteur succeeds in establishing principles which are apparently very simple and unambitious only after lengthy gropings, ranging over fields which at that time were as separate as crystallography, structural chemistry, the transformation of carbohydrates, botany, fermentation, medicine, etc. Any over-simplification would be, to say the least of it, fallacious; suffice it therefore to record that the notion of organization, studied in depth and refined by molecular biology, is an acquisition of experimental biology. In the second half of the nineteenth century, organization is seen no longer as the starting point but as the goal of a knowledge of life, not as that which is taken for granted by explanation, but as that which is to be explained. A knowledge of life becomes a knowledge of organization.

But this knowledge is scattered over various fields. When Mendelian genetics developed, this development was independent of chemistry. Admittedly, Mendel no longer thought in terms of the race or the individual, but of character. Where is the site of character? Where is the heredity factor to be found? To discover this, further developments were necessary. Above all, it was necessary to find experimental material for cross-breeding other than strains of peas, rats, guinea-pigs or drosophilae. But in biology, the choice of the material for study is frequently a discovery in itself. This was the case with bacteria.

Cell theory satisfactorily unified the sphere of life, but this sphere was one from which bacteria were excluded. Elsewhere, discoveries were piling up in relation to physiological activity and heredity, but there was nothing which made it possible to co-ordinate all the facts into a single theory. Biology explained life piecemeal. What was needed was to extend the limits of various sciences, and to find a new link between disciplines. At this stage a chemist and a geneticist came together to study mildew; they discovered that an organism's heredity governs its chemistry, and that between gene and character there is the action of protein. The reactions of the metabolism became the subject of study for genetics, which was thus led to take an interest in bacteria.

Thus heredity extends to include bacteria; but this actually makes things simpler, since with an organism as rudimentary as a bacterium, the various analytical techniques become so simplified that they can be employed simultaneously. The very change of scale in itself leads to the chemistry of genetic material. The work of Sutton, Boveri, Morgan, Muller and others had already made it possible to pinpoint the location of heredity factors. A knowledge of the deep inner nature of genetic material, of the chemical composition of the chromosome, then became the lynch-pin of molecular biology.

It was not possible for this to take place without starting from everyday experience, bearing in mind the usefulness of breeders and farmers, in particular farmers, who were the only people capable of producing populations in adequate numbers for the geneticist's purposes.

Thus experiments on micro-organisms, bacteria and viruses made it possible to unify the genetic and chemical approaches. From that point onwards the two types of studies went hand-in-hand: the characters of the bacterial cell, their variations and recombinations; and the characteristics, corresponding proteins, structure and function of genetic material. Lederberg and Tatum had blazed a trail by their work on strains of the *Escherichia coli* colibacillus, which highlighted genetic recombination. Avery, McLeod and McCarthy identified the presence of deoxyribonucleic acid (DNA) at the outset of all genetic transformations and mutations. And the discovery of structure established the chemical nature of genetic information. The properties of the cell were then seen to be the output, as it were, of the programme of a living computer.

But the observation of spontaneous mutations leads naturally to bringing about deliberately contrived ones, and it is these which have made it possible to establish that the genetic message is carried in DNA, the substance responsible for the transfer of the genetic characteristics of the donor bacteria to the receiving bacteria.

Deliberately produced mutations enable us to act on cells without destroying them. This ability represents a new type of experiment: molecular biology brings about 'from within', and as a consequence of mutations, the modifications which experimental physiology—when it did not confine itself merely to the observation of pathological phenomena—brought about in the organism 'from the outside', either by mechanical intervention or by means of toxic substances. Mutation creates the abnormality which makes it possible to analyse natural deviations. One might say that deliberately produced mutations are the work of an 'experimental demiurge', using the microscopic mechanism of the enzymes to realize his vision in the living material of life itself.

Here we come to the second line of approach leading up to this book, that which considers the limits to be placed on the powers conferred on man by this new knowledge, and invites us to reflect on the moral discourse to which scientists should pay heed.

In point of fact, this is not the first time in the history of biology that the moral problem has arisen, to what extent experimentation respects ethical principles. The protests levelled in the nineteenth century against experiments on animals were inspired, *mutatis mutandis*, by the same concern. Thus vivisection was accused of 'deplorable lack of concern, which allows young people to contract habits of cruelty which are as radically detrimental to their moral development as they are completely useless, to say the least, for their intellectual development' (A. Comte). Yet the practice of excising organs or cutting off the heads of animals goes back to the mists of time; it was only when it was performed in clean and peaceful laboratory surroundings that it suddenly caused an outcry. Anti-vivisection organizations were founded, and it is amusing to note that the wife and daugher of the French physiologist Claude Bernard were among the foremost militant anti-vivisectionists of their time. Their efforts were moreover not in vain, since they resulted in the framing of regulations for the sale and purchase of laboratory animals. In the United States, Congress has recently adopted legislation on these lines, following the example of countries as varied as Canada, India and the United Republic of Tanzania.

However, the ethical problem posed by molecular

biology is undoubtedly more serious than that raised by vivisection. Here experimentation involves not only, as in the past, already-formed structures, but also the programme which governs them. The selective pressure exerted in a laboratory on a population of bacteria can even become so effective that one can obtain almost at will monsters in which the selected function is impaired by a mutation. Such impairment of the programmes of bacteria and viruses might well turn out to be disastrous. Between this, and saying that genetics is to biology what nuclear physics is to physics, there is only one step, which is frequently taken. Unlike anti-vivisectionism, which sought to protect animals from man, we are now, astonishingly enough, trying to protect man from the bacteria which might possibly escape from laboratories. Techniques for isolating and recombining DNA segments make it possible to construct *in vitro* DNA molecules with unforeseeable or uncontrollable biological activities which may, in certain cases, be pathogenic; and it is feared that the pathogenic element may be disseminated among human, animal or plant populations.

We thus see genetics going the same way as atomic physics, and offering equally fertile grounds for debate on the relations between science and society. The first public discussion on the subject was at the 1973 Gordon Research Conference on nucleic acids. Invited to examine the problem, the United States National Academy of Sciences had appointed a group of scientists, who supported Paul Berg in recommending (in 1974) the cessation of certain types of experiments pending the development of methods which would prevent any risk of the dissemination of artificially created recombining molecules. They also asked that an advisory committee be set up to assess the real dangers inherent in manipulations of this kind, to develop operative methods which would reduce to a minimum the dissemination of dangerous or suspect molecules, and to draw up rules for researchers to observe in the future.

The celebrated conference held in Asilomar (California), in February 1975, authorized the resumption of these experiments, but laid down stringent safety requirements, both for the laboratories in which such investigations are carried out and for the biological systems studied: special protective measures against the risk of contamination are now in force (at the Institut Pasteur, in Paris, a special genetic unit is being constructed on purpose), and it is recommended that types of bacilli should be studied which cannot multiply in man.

Recently a consensus appears to have been established on the need, not to control science or to limit experiments, but to draw up safety standards for research into genetic recombinations *in vitro*. This concerns measures to obviate the potential risks involved in experiments (the reader will remember cases of tragic accidents to laboratory researchers and technicians working on the production of vaccines, for example those who contracted a fatal disease after manipulating tissues which carried latent monkey B viruses in the herpes group). There is no doubt that a study should be made of the various systems of regulations already prepared, or being prepared, not only in order to clarify the relations between science and deontological practice within each discipline, but also to devise a deonotolgy which can be applied internationally, something which the existence of national systems of regulations makes it possible to envisage.

Obviously, these considerations have not left Unesco unconcerned, since science, and the problems it raises in the world today, form part of its programme of reflection and work. From 24 to 27 June 1975, Unesco held at Varna, in Bulgaria, a meeting which is at the origin of the present publication, which consists of a review of the contributions made by eminent biologists and philosophers to that meeting. The subject-matter is set out in an ordered and considered form, faithfully reflecting the state of the question, which continues to develop; it is a review which occasionally ventures on new ground, when the author outlines, in an entirely personal way, the broad lines of a philosophy of life.

Following on from the Varna meeting, this book marks a new stage in Unesco's consideration of the problem, or rather in the history of Unesco's contribution towards clarifying it.

Previously Unesco had organized in Madrid, in collaboration with the Spanish National Council for Scientific Research, a symposium dealing with the problems and positive results of scientific research into molecular genetics. This symposium examined various possibilities of genetic manipulations applied to the solution of vital problems of concern to mankind. It drew attention to the accelerated pace of genetic erosion, and suggested the establishment of an international network of 'seed banks' which would make it possible to preserve genetic resources, in particular plant resources. It stressed the need to re-think the idea of the 'green revolution', placing it in the context of a new international order.

On the delicate question of the control of research, Unesco is continuing its analytical studies by encouraging increasingly detailed discussion of its most clearly defined aspects. In a sense, however, the Madrid meeting marked a turning-point in this respect, with Professor Roman de Vicente, a member of the Spanish National Council for Scientific Research, stressing the need to develop the research which has made it possible to transfer and reconstitute the genes of rat insulin with the *Escherichia coli* bacterium. It is more than possible that this research represents a preliminary step towards the development of a unified view of living beings. The era ushered in by the discovery of the role of DNA as the carrier of genetic information has not exhausted all its resources. Experimental investigation continues. When the first secret has just been wrested from a complete nucleotidic sequence of the viral genome, that is to say its complete genetic message, there are grounds for hoping that we shall in the near future establish the DNA sequence of systems other than viruses. Some of the participants in the Madrid symposium stressed that the bacteria and the viruses used were no more dangerous than ordinary bacteria and viruses. The possibilities of genetic recombination which have enabled geneticists, as noted above, to obtain viral subtypes similar to those of the influenza epidemic of 1918–19 will also enable us to extend our knowledge of illness and the soil on which it develops.

It is for these reasons that the approach to the moral problem is partly taking the form of an investigation bearing both on measures to guarantee that research will be continued in the best possible safety conditions, and on the content of a research policy which would free mankind from the fear of the bacteriological armamentarium. Whether this is possible or not, only the future can say. At least the project gives its full significance to an exacting philosophical reflection on what should be the role of knowledge and its relation to action, on how to develop a world in which man can live better—concerns which loom large in the current problem of civilization, and in the malaise which our age must overcome.

Contents

B

Foreword

For a long time now, and in many countries, symposia, seminars and study groups have been grappling with the ethical implications of the advances made in the various life sciences. Attention is directed above all to a number of deontological problems (defining a 'code' for research, and its application), and to the moral imperatives which must be respected in making use of the possibilities opened up by these advances.[1] Today, reflection on this subject is still undoubtedly essential. It should, however, be made both more radical and more extensive in scope: more radical, since the deeper our knowledge goes the more we are obliged to re-examine the basic tenets of the science of man, indeed our very conception (including our actual definitions) of life and death;[2] more extensive, since the purview of biology has grown considerably wider, in relation to both the individual and society, both now and for the future. This being the case, there is manifestly an urgent need to formulate a genuine policy with regard to life, itself related to a coherent ethic.

Mindful of this need, Unesco convened a meeting of biologists and moralists at the Centre of the Union of Scientific Workers in Varna (Bulgaria) from 24 to 27 June 1975.[3] The biologists reported on the state of current research in their various disciplines, and indicated as far as was in their power the implications of such research for the world today—incidentally giving evidence of a profound desire to safeguard the rights and integrity of the human individual, and of an acute sense of their responsibility towards society, together with concern for the use which might be made of their

work for political and economic ends. The moralists
associated themselves fully with what had been said in
these respects. A number of views were then expressed
with regard to such varied and burning issues as gene
therapy, the regulation of human behaviour, organ
transplantation, the prolongation of life, etc. However,
owing to the evident need to carry reflection on the
subject a stage further, it was agreed to try and establish
some kind of guideline for a combined approach to the
problems involved, and to base the elements of a reply
to the questions raised on something other than a
merely pragmatic attitude. It is true that with these
problems there is very often 'no universal solution'
capable of reconciling requirements and values which
sometimes appear to be conflicting, e.g. to mention
only one such area, the vital interests of the individual
and of society.[4] This being the case, the final report and
the recommendations made to the Director-General of
Unesco stressed in their turn the 'urgency of developing
a new ethic'[5] and the necessity to 'continue and
strengthen the programme on science and ethics
through keeping the ever-changing subject of ethics in
relation to the sciences under constant review . . .'.[6]

It was as a contribution towards meeting this
twofold need that we were asked to present the main
ethical conclusions emerging from the work begun at
Varna. In response to this invitation we might have
prepared a kind of abstract of the reports and discussions. We thought it preferable to take the bolder
course of seeking to present the material submitted to
us in an organized form, adopting from the outset the
experts' second recommendation: 'help define the
concept of a new functional ethic and encourage
research in ethics'.[7] We are then in duty bound to draw
the reader's attention to four comments:

1. To meet the concern expressed by the term 'functional ethics', the basis adopted for this study will
be the scientific data as they were presented in the
course of this symposium. Clearly, however, it is in
no sense our intention to describe the entire range
of possibilities available today, for instance the
efforts that are being made to induce desired genetic
information *in vivo,* or the techniques used to
produce recombining molecules. Not being a
biologist himself, the author is in no way competent

to do so (in any case, since it was the whole field in which biologists are currently working that was evoked at Varna, any account of it would not only run to several books, but would require the assistance of a host of experts). Thus our object in basing this study on the scientists' contributions is purely to elucidate or illustrate the ethical questions brought up by them.[8]

2. Nor are we taking it on ourselves to outline any kind of 'new ethic' (we shall demonstrate that such an undertaking would require extensive interdisciplinary co-ordination). Our intention is limited to an attempt to indicate the direction in which, in our view, 'research in ethics' relating to biology should be guided, or, on still more modest lines, what might be the starting point for further exchanges between scientists and moralists.

3. In other words, this study does not consist of an account of the discussions which took place at Varna. Rather does it take the form of a reflection on the body of material thus constituted.[9] Our objective would be attained if, through the criticism and reactions elicited, we prompted a more searching investigation of the relations between biology and ethics, an investigation which is all the more necessary and urgent in that, in view of the progress achieved and the possibilities opened up, these relations most undeniably raise one of the major problems of our time.

4. For the same reasons, this tentative review is the sole responsibility of its author, and in no way commits the participants in the symposium, even in cases where reference is made to the papers they presented.[10]

Notes

1. By way of a reminder, it will be recalled that 'deontology' denotes the theory of the moral duties inherent in a profession; by 'morality' is meant the set of rules of conduct considered as necessary in order to respect the good; 'ethics' is the science which seeks to determine what good is. By analogy, it can be said that ethics is to morality what biology is to medicine.

2. For example, since 1970 there have been a great number of philosophical essays written by biologists, using their knowledge to extrapolate, as it were, a global conception of man.

3. The list of participants is annexed. As will be seen from their titles and qualifications, the word 'biology' was understood at Varna in a very broad sense, as also in this study, to cover all the scientific disciplines which relate directly to life.
4. This sentence summarizes one of the points elaborated at greater length in the final report.
5. Final report of the Varna symposium.
6. ibid., first recommendation.
7. ibid., second recommendation.
8. Although the papers presented at Varna took account of the most recent trends in research and technology, it is possible that since then new fields of application have been more fully explored or delimited. However, any such advances in knowledge cannot possibly be of such a kind as to modify the ethical issues facing scientists or affect the conceptual approach emerging from the present study.
9. We might adduce in support of this study a large number of references to scientific articles or publications which bear out the statement by participants or our own reflections. We have not done so, in order to avoid distracting the reader's attention by a plethora of notes.
10. As it is not possible to quote all the reports *in extenso* (a total of approximately 1,000 typed pages) the paragraphs we consider the most significant are inserted in the margin, linking them, by means of the signs *, †, to the phrase or sentence in the work which they amplify or illustrate.

Problems
and implications

In order to see the questions raised in broad perspective, and also to appreciate the need for a 'new code of ethics' and the conditions that should be fulfilled if we are to formulate such a code, we must, in the first place, attempt to sum up the nature of what is involved.

Biology as an integrated science

In recent years, 'new basic principles in biology have led to an advance, comparable to that of physics and electronics during the past few decades . . .'.[1] Another fact, rarely analysed in its own right despite the importance of its implications, is that biologists have been forging closer links with their research colleagues in other scientific fields, whose discoveries and techniques have contributed considerably to advances in the various life sciences. Until recently biology could be regarded virtually as a self-contained science. This is no longer the case; today it is far more closely integrated with the whole body of scientific knowledge.

The primary indication of such integration is, obviously, an interknitting of all the various life sciences. To give a single example, organ transplant surgery cannot disregard the basic principles of molecular biology, if only because they enable it to deal successfully with rejection. But biology interrelates in an even more spectacular fashion with other sciences, whose theories and techniques it systematically exploits. It is hardly necessary to recall the extent to which contemporary biology is indebted to advances in physics (such as the laser and radioactivity, which make such an important contribution to medicine and research) and, even more so, in chemistry, which has become a key science for anyone wishing to understand vital phenomena, and plays a constantly increasing part in the treatment of a wide range of diseases. Conversely, it is evident that advances in biology have had considerable repercussions in these other scientific fields, if only by providing an incentive to further research.

A further instance of this integration is the exchange of scientific 'models'. It is not simply that similar research techniques tend to be applied in the various sciences; there is also a concerted effort to discover areas of structural convergence. Thus, the 'logic' of the living being is often expressed in language borrowed from data processing. Conversely, many a sociologist would apply this 'logic' to his reflections on society.

Equally significant are the links woven between biological research (particularly biochemical research) and the spheres of agriculture, industry and even weapons manufacture. By the same token, biological research comes under pressure from powerful interests, which, in their impatience to see results, impart their own slant to the principles of professional ethics by which scientists are actuated and the purpose of their research.[2]

To an increasingly large extent, the fate of mankind depends on the advances made in the life sciences. This is obvious whether we consider: (a) the struggle against the dangers from pollution of all kinds, and the preservation of equilibrium in the ecosystems, and even the biosphere; (b) agricultural development, since agricultural output must be increased rapidly if disastrous food shortages are to be avoided; (c) the preservation of the genetic endowment of mankind and the protection of health, which is affected on a world scale by multiple forms of exchange, transfer and intermingling among individuals and populations;[3] (d) the need for measures to counteract the mental traumatisms or stress suffered by a growing number of individuals, who are incapable of coping with the manifold tensions of life in the megalopolis and the numerous forms of conditioning to which they are subjected; (e) the actual structures of the various societies whose make-up it is becoming possible to alter (if only through widespread contraception or the availability of sex choice); (f) naturally, the control of world population growth.

Many more examples could be adduced, but what has been said suffices to indicate the extent to which the life sciences are henceforth bound up (by increasingly close and complex ties, through action and reaction) with the totality of human becoming, taken in all its dimensions, which range from the physiological and psychological to those of a cultural, social, economic and political nature.

Basic areas of ignorance

The pressures and demands weighing upon biology today give rise to numerous theoretical and practical extrapolations which tend to treat incomplete discoveries as if they were global and definitive and hasten on to exploit the possibilities held out by such discoveries. However, there is no denying the fact that these extrapolations and practical applications take place against a background of persistent areas of basic uncertainty, and even ignorance.

It is common knowledge that while scientists are in overall agreement as to the fact that life has evolved over millenniums, they differ when it comes to offering an explanation of the mechanism(s) involved in such evolution. The debate between those who believe in a hormic principle and the supporters of the pure-chance theories is still very much alive (even if the latter seem to have had the best of the argument in recent decades, particularly in view of the advances in biochemistry). It is not simply that the solution to this problem influences our conception of man and the universe; it also gives a direction to biological research as a whole. Nor is it possible simply to 'bracket' this problem when engaged in the practical application of scientific knowledge. It is clear, for example, that we are trying to bring about increasingly profound mutations in living beings (plants, animals and even human beings) without adequately understanding the relationship between mutations and evolution, or even between what is reversible and what is irreversible.

In the same way, we have a very limited knowledge

of the laws governing the living being (considered according to its internal structure and its relationship with the environment), of its constituent parts and of its actual or potential capacity in terms of organization and adaptability. Accordingly, many practical applications of biological knowledge have had—or can have, since the effects are often delayed—injurious repercussions. This has already occurred in the case of the ecosystems and the biosphere. However, the interventions which we are preparing to carry out, particularly in genetics, will have far more extensive implications yet, especially if the use of viruses as mutagenic vectors is developed.[4]

In the case of man, the unknown factors are even more numerous. Our knowledge is deficient in regard to the physiology of the brain (what is the exact function of frontal lobes, the development of which is peculiar to man?), its activity, its dependency and its action in relation to the basal metabolisms, the emergence of mental activity in its various forms and its role in the internal regulation of the organism as a whole. At best we understand certain effects, certain functions. This is an area, however, where most of the mechanisms and the causes involved are still a mystery to us.[5]

It follows that we are to a large extent incapable of forecasting or controlling the physiological and mental, individual and social 'fallout' of certain practical applications of scientific knowledge, and this at a time when, as we have seen, biology is faced with increasingly urgent pressures and demands from outside, which remove it further and further from its natural bent.

Partial efficacy and global consequences

This process has reached such proportions that the entire relationship between man and biology is substantially modified. Whereas, formerly, we confined ourselves to 'letting be', coping as best we could with whatever befell, contemporary culture is concerned with 'bringing into being' and has growing aspirations to master the essentials of life. Such a transformation would be positively exciting if we did not feel bound to rush from incomplete knowledge to action; and if it were not for the growing tendency to rely exclusively on scientific and technical achievements, as if they were in themselves capable of obviating every error or instance of recklessness, and even of reversing what is irreversible.*

The basic criterion of action and of the practical application of scientific knowledge has become the fact of something being possible. But 'just because hitherto unimaginable developments are possible ought we to try them?'[6] This question is all the more crucial as we are not fully in control of the outcome of our scientific activity; the possibilities available to us are more or less sporadic, and we are unaware of their long-term effects.

These possibilities are becoming more and more numerous, as may be seen from the list of subjects considered at Varna (and others which could have been included on the agenda), not speaking of the secondary questions which emerged from these subjects: (a) genetic intervention: gene therapy, cloning (the production of identical copies of a living being), the treatment of genetic disease; (b) new techniques of conception: genetic counselling, *in-vitro* fertilization and embryo research, intra-uterine diagnosis and the termination of pregnancy, sex choice and artificial insemination; (c) family planning and population control; (d) in regard to the fully formed individual: organ transplants, biological brain research, and behaviour regulation; (e) senescence, the prolongation of life, euthanasia and death.

Thus, man's whole life is affected by an increasingly complex assortment of discrete possibilities, which do not assume any real kind of organic unity. No doubt this is to a certain extent inevitable since, while research can be controlled, scientific discovery (and, consequently, the possibilities which it makes available) is due in part to genius and even to luck, just as unscheduled demands or pressures may force scientists to carry out their work with excessive haste. The fact remains that this compartmentalization of the possibilities made available by scientific discovery poses a very serious problem since each of them is called upon to have lasting, if not irreversible, effects on individuals and society.

We cannot bear too much in mind that for contemporary biology and medicine alike, human 'nature' is relativized and health is no longer 'something to be

* Culturally at least, health problems tend to be posed in terms of *ability to do* rather than in terms of *doing*, in a kind of dialectic between competence and incompetence, as well as between what is possible and what is not possible (at present). By the same token, responsibility, in ethical terms, is geared to this ability to do, with an increasingly marked tendency to disregard any other frame of reference.— *From the author's paper.*

taken for granted'. The tendency is to try to restore balance to life. But such a balance is a question of organicity. The danger of discrete, compartmentalized applications of scientific discoveries is therefore obvious. This danger is well known by doctors who are often compelled to carry out treatment purely in the—often vain—attempt to offset the effects of another form of treatment, one form of treatment leading to another in a relentless process which frequently becomes intolerable. It would be easy to demonstrate that all the possibilities held out by the discoveries of biological science involve generally unforeseeable compensation processes (which can only accelerate). It could also be shown without difficulty that all these advances have long-term global repercussions. Thus, we have mindlessly (How else?) disturbed numerous ecosystems. Or certain practical applications of scientific knowledge, which may be tolerable if not justifiable in terms of individuals, have altered this or that element in the basic make-up of mankind. A typical instance here is the decline in mortality, particularly infant mortality.[7] This is obviously one of the main factors in the incredible world population growth, which has resulted in the need for widespread birth control. However, it has also accelerated urbanization, which in turn seriously modifies the psychosomatic conditioning of individuals, not to mention social, economic, cultural and political structures.

By way of example, and since it is possible, *grosso modo*, to determine its various effects, let us consider the practical implications of artificial insemination. The techniques of artifiical insemination have been perfected in animal husbandry with a view to obtaining a higher output and improved breeds. Today human couples resort to artificial insemination as an answer to certain forms of infertility or accidental impotence, and sometimes a donor is involved. It is 'used for therapeutic purposes in cases of male infertility or in the presence of certain genetic diseases in the family'.[8] Taken in themselves the questions raised by artificial insemination with donor sperm (AID) may appear to differ little, at least in theoretical terms, from the various questions raised by adoption. In reality, manifold psychological problems result from this form of artificial insemination. Father and mother do not have the same type of kinship with the child: in the full sense of the word, only the mother is the child's parent and the father is required to adopt it. Another point is that the child is the permanent testimony of the man's infertility. In the event of marriage problems or a crisis of authority between father and son, could not this unequal status drastically aggravate the conflict?[9] It is for this reason that in a numbr of countries AID is used with extreme caution and future parents are urged to give serious consideration to all the implications of their decision.[10]

Apart from the various interpersonal implications, AID would appear to have socio-cultural (and political) repercussions.[11] It is desirable to restrict the use of this technique to a limited number of couples (only those who are disturbed by the infertility of the husband). However, the problem will become more complicated when, as the inevitable outcome of artificial insemination, sperm banks are widely established and sex choice becomes available.* 'We know that in most societies of our planet, and the developing countries are no exception to this, significantly more males would be selected than females.'[12] However, sex choice may have undeniable advantages for society (if the family 'pattern' corresponds to two children, as is the trend in most developed countries) or for the parents (for example, when they want a boy at all costs and already have a number of girls). It would be possible to control sex choice by various means such as making the technique available for the first child only, or by public discussion of practices if serious deviations are noted in the male/female sex ratio of roughly 1.05 which represents the balance established by nature.[13]

The truth is that such considerations are valid in terms of the family structure as we know it. However, significant changes have been made in several countries in legislation concerning illegitimate children and the status of unmarried mothers. What would happen, then, if the current 'don't-get-married' trend of the younger generation were to acquire greater prevalence? Would a young woman or, putting it more dramatically, a neglected wife have the right to resort to a sperm bank in order to have a child of a certain sex? The social upheaval which would result, if this were possible, can be easily imagined. It would occur on a cultural plane (an accentuation of the present trend to dissociate

* Research in isolating human X- or Y-chromosome bearing sperm is yielding promising results. Once a reliable separation is achieved, sex choice by the use of artificial insemination would be feasible and this method could become part of family planning . . . .— *Final Report.*

fertility and sexuality), and on a social plane (a change in the status of women, in basic parental and social patterns, in population structure, etc.). At the moment, in a number of countries where sperm banks exist, controls are generally strict (although not embodied in legislation): artificial insemination is only used when both parents give their consent, and this form of intervention is still made available only occasionally. But if this practice were assimilated culturally, would it be possible to avoid extending it to other cases? Already the question is being asked openly: to what extent is it legitimate to deny a married or single woman the chance of motherhood? In this respect, what are the rights of the child? And those of society?

What kind of ethics?

We have dwelt on this example at some length[14] since it is possible to gauge at least the major cultural, social and demographic implications of artificial insemination. A basic truth emerges: as soon as we interfere in one way or another with life, the repercussions are global. They concern the physiology and psychology of the individual, social structures, and even ecosystems. These effects may be beneficial and/or harmful.

Accordingly, there tends to be considerable disagreement on the question of a moral approach to these problems. A growing number of biologists and even ethicists refuse to make a value judgment which would be founded upon an abstract or arbitrary generalization or would be likely to involve such a generalization being made. Their position is that proper account should be taken of the diversity of individual and collective (cultural, social, economic and political) situations.* It is based, furthermore, on the fact that other referents which until recently seemed to be universally accepted and permanent, such as 'nature' and 'life', are undergoing changes of meaning, which arise partly from a better understanding of the constituent mechanisms of life, but are also a result of the more successful integration of biology with the corpus of physiochemical sciences, and also with the socio-economic and socio-cultural spheres. Lastly, this refusal to lay down abstract general principles is an

* Although there was a lack of consensus as to what 'ethics' means today, the general feeling was that ethical standards and value systems concerning men are dynamic and in constant change, and that moral approaches and value systems undergo changes with time and according to specific geographic areas, ethnic groups, societies and culture settings. The attraction of a certain absolute character of values remains but its dynamism and its expression through the values can be modified. Hence we cannot reason deductively from *a priori* fixed and determined rules about the ethically justified nature of scientific action. Thus the majority suggested that a functional ethics directed towards man's present and future happiness and his biological fitness be sought, founded on man's privileged position in nature, on his autonomy, as well as his dependence. As such, rather than being universal and destined for global application, it should be flexible and adaptable to the needs of different population groups in various regions of the world, or as one participant put it, 'new occasions teach new duties'.—*Final Report, echoing the paper presented by Dr Vasken M. der Kaloustian.*

indication that the solution to the problems raised 'would have to incorporate parameters which appear to be mutually opposed, such as present/future, certainty/ risk, power/powerlessness, freedom/constraint, or to go further, collective and individual, personal and social, well-being and ecosystem, etc.[15]

Can we, however, content ourselves with a purely situational morality? Would we not be liable to remain 'tied to a purely relativistic outlook which may lead us astray?[16] We have seen that we could be led astray by the widespread use of various techniques made possible by biological advance. While they may be justifiable in individual terms, their general application would bring about qualitative changes in the nature of society and give rise to imbalances or undesirable chain reactions. We may also be led astray in consequence of the various economic, social, cultural or political pressures to which biological research—pure or applied —is exposed.

A number of crucial questions may be asked in this context: 'It is good we do everything we can do technically? Is it better to do nothing until we can do everything? How can we decide what we can responsibly do as opposed to what is technically possible?'[17] Any consideration of these problems inevitably involves asking a further question: What are the aims of biological research and its various practical applications? That human life should benefit or be improved or served? Then it is necessary 'to define as clearly as possible what meaning is to be assigned to words like "benefit" or "improved" or "served"; in other words: "beneficial to *whom*", improve *what*", "serve *whom*": is it man as a biological entity or man as a human being who is not solely and simply defined by his biological set-up'.[18]

This is clearly a crucial point: biologists should continually remind ethicists in particular that man cannot be dissociated from his somatic features nor, therefore, from the biosphere as a whole. Conversely, however, ethicists, along with others, ought to bring moderation to bear on the projects of biologists and remind them that man is more than the sum of his somatic features.[19] The following two observations are essential to an understanding of our theme:[20]

The analysis of the relationship between 'biology' and

'ethics' is but one facet or stage of the task involved in establishing the coherent ethics which any biologist in his senses would unreservedly welcome. Such an ethics would presuppose a complex, 'interdisciplinary' and 'transdisciplinary' approach, or rather the elaboration of genuinely scientific systems, in order to comprehend the infinite diversity of the relations which comprise the human being.

In the endeavour to elaborate such systems, care must be taken to ensure that each discipline makes an adequate contribution and that 'models' are not integrated to such an extent that the specific features of the various sciences are forfeited. The generalization of certain quantitative approaches or the combination of, for example, the biological, psychological and sociological aspects involves the risk of misinterpreting human nature. (In this context, it is surely important today to give not only fuller consideration but also greater weight to the research of ethicists and philosophers, as well as to the axiological approach to the question of the destiny of mankind.)*

To put it another way, it is important, in our study of the relations between biology and ethics, to beware of extrapolating unduly from biology, as if the latter could serve as the frame of reference for all knowledge or could embody all science. At the same time, we must endeavour to understand how biology defines or represents the common good and individual good, before considering how this question has necessarily to be viewed from the standpoint of other sciences or other disciplines.

This would appear to be all the more necessary since, as we have seen, ethical judgments should not be formulated arbitrarily and in an abstract manner; nor should they emanate from the outside in the form of approbation or condemnation after the event. The effects of science, it must be realized, are reaching out further and further, in both spatial and temporal terms.† Can biology 'resign its responsibility into the hands of non-scientific authorities—political, economic or social? Can it expect moralists simply to label the applications of scientific discoveries as good or bad, giving "directions for use"?'[21]

'May we not ask ourselves whether science and

* We cannot endorse the view that the scientific and axiological approaches are mutually exclusive. Neither scientism nor anthropological humanism can supply us with a correct solution to the problems concerning man . . . . The axiological and the scientific, the normative and the factological are two aspects of one single view of the world . . . . —*Extract from the paper by Professor Stéphane Anguelov.*

† The new powers of science have a virtually unlimited range of action extending through the whole of nature in general and covering the sector of life in particular. . . . Today nature is seen to be vulnerable to the technological assault which threatens to impair its integrity . . . . Man can be subjected to changes in his biological and psychological make-up, and this adventurous road is fraught with dangers to the essence of his being. . . . [Until recently] the results of human action were directly linked with the causal agent. . . . Man could gauge its consequences and adjust his behaviour in the light of his personal experience. Today the causal agent is separable from the effects of its action by generations of living beings. . . . A factor which is common to these three new characteristics of science is the irreversibility of the damage caused by man, since no rehabilitation can be envisaged—and even if it could be, it would not affect the same

organisms—particularly when living beings are involved and, more specifically, the human race.—*Extract from the introductory paper by René Habachi.*

* If we are to be in the vanguard of the movement concerned with (the problems raised by the advance of biology), we should, perhaps, resolve to adopt a dual approach encompassing the whole issue:
(a) In a world where there is a growing sense of solidarity and more and more evidence of the interdependence of all its parts, we should adopt an approach that embraces all cultures. ... Here, then, there is a *responsibility to respect the principle of solidarity.*
(b) In the case of a science which is still developing, we should make an effort to visualize the limits of its development, not from the standpoint of the scientific content, the progress of which is fairly difficult to forecast, but from the point of view of its global power which affects— and will increasingly affect—the whole of nature and mankind... Here, then, is a *responsibility to respect the principle of the totality of man.— From the introdcutory paper by René Habachi.*

c

ethics can be considered today as two *independent* spheres which can only exercise an external form of mutual supervision? Ethics, on being challenged or merely questioned by science, is then constrained to act as a brake on the applications of scientific knowledge, so that its intervention comes too late and in a situation made more difficult by the delay. May we not ask ourselves, on the other hand, whether the two spheres of science and ethics should not be envisaged as closely *bound together*? The research work of the true scientist would then bear simultaneously on ethical issues and, although science would not identify with ethics, the effect of this approach would be to establish that it is at the very source of scientific research that the ethical intention should be present in other words, in the attitude of the scientist and of scientific institutions, as also in scientific forecasting with regard to the use which is likely to be made of the achievements of science.'[22]

Thus, while we recognize the necessity of taking into account the specific situation of each living being and comparing biological data with other scientific knowledge, it would appear that there is a 'duty to integrate' the ethical factor into scientific research as such, not for the purpose of restricting the freedom of the mind to pursue the path of discovery, but in order to alert the scientist to the need to use his freedom to serve the common good of mankind.[23] Conversely, however, ethics should incorporate biological data and its serious implications in the matter of respect for life.[24] Would not this dual requirement best be met, at least in the initial stage, by the introduction of some form of basic code which would provide the framework for understanding the nature of the 'good' in relation to biological research and its practical applications? However, this basic code would also have to satisfy various other conditions:

It should be possible for it to be applied to the whole of mankind, irrespective of culture or level of development—i.e. the social structures—while it should at the same time respect the specific characteristics of individuals and peoples.[25]*

This basic code should also be sufficiently flexible, if not resolve, at least to decide between, and certainly to situate better, the previously mentioned contradictory demands arising from parameters such as individual/collective, personal/social, freedom/constraint, quantitative/qualitative, politico-economic/cultural, power/powerlessness, present/future, etc.

In more rigorous terms, it should establish a frame of reference that would make it possible to define (in the sense of specifying, determining and delimiting) the 'projet' or project we pursue. At present, this project is designed primarily to eliminate chance or the accidental (particularly disease), to preserve or restore the symbiosis of living beings (compensating for deficiencies) with the environment which we re-create, and the symbiosis of individuals with the various structures which we have installed (principally industralization and urbanization), and to satisfy new aspirations and demands. Such a project rests largely on utilitarian considerations which present themselves to us in a more or less imperative form. However, in the matter of life, the utilitarian approach often involves opposing interests: that is obvious if one is considering the needs of each part of the body (what is seen to be good for one part could be harmful to another); likewise, personal 'good' and social 'good' do not necessarily coincide. In other words, the useful has to be regulated and tempered by the organic.

If there is a project or design for the future of mankind, it must discern what is necessary (something which includes yet extends beyond 'what is useful'), and the more artificial are the types of intervention proposed, the more rigorous must be its criteria in this respect. It is not sufficient to state that it is important, in this context, to know and respect the primordial functions governing survival: conservation, adaptation, reproduction, the emergence of qualitatively higher forms of existence. It is also necessary to avoid 'dis-integrating' the living being by satisfying this or that aspiration or by treating this or that disease, with considerable use of specialized techniques, without any regard for the resulting imbalance or the overall personality. Nor must the living being be 'de-historicized', i.e. cut off completely from its biological and socio-cultural past and future. Our emphasis upon '*what is necessary'* imposes a positive duty to forecast, integrate and relate to an end.

It is true that these various requirements are difficult, if not impossible, to observe, since we are unaware of the effects of our actions, at least to a certain degree, their repercussions being remote in spatial and temporal terms. There can accordingly be no question of adopting a basic code which would take the guise of a new form of determinism. Rather, it is a matter of establishing certain guidelines which would serve as points of reference enabling us to situate, consolidate and constantly modify our 'project' at the same time as we act upon living beings and attempt to arrogate to ourselves the right to control life.

This type of ethical approach, in the form of a code setting out certain guidelines, not abstract yet having universal validity, capable of transcending the mutually opposing demands of biological science and regulating 'what is useful' in terms of 'what is necessary'—does it not imply that one has already entered, and enters ever more deeply, into the 'logic' behind life?

This is what we shall endeavour to do in the following chapters, presenting in this context the main conclusions or recommendations put forward at the Varna meeting. It is clear, however, that what emerges from our endeavour must be treated with caution, for three main reasons:

The first concerns the incomplete nature of the endeavour: we can refer here only to certain elements of this 'logic', and shall try to show their 'value' and their application in terms of ethics.

Another factor is the transitory nature of such an endeavour: it is clear that an understanding of this 'logic' of the living being depends upon the state of our knowledge—that of the author, who is not a biologist and whose competence in the matter is limited, and that of the professional scientists, who have to keep abreast of scientific advances and overcome the basic areas of ignorance which we have underlined. Taking into account this ignorance (our own and that of the scientists) is it not untimely, if not futile, to embark upon such a study? The answer is: 'No'. On the one hand, we shall confine ourselves to a number of elementary yet basic and uncontestable considerations. On the other hand, our aim is to make a first step, to urge the reader to enter into this 'logic', rather than to establish any norms.

Lastly, the third limitation of this work is its inherently biased character, due to the fact that it is mono-disciplinary or, at the most, bi-disciplinary whereas, directed as it is to man, any ethical approach must necessarily be transdisciplinary, as we have seen. Our intention, therefore, is not to extract from biology a blueprint, design or 'project' for society or to determine the goals man should set himself in his search for the 'good'. Rather, it is at most to specify the conditions under which he can hope to realize such a blueprint or such goals. However, directed as ethics is to life—since man is essentially a living being—an ethical approach clearly entails observance of these conditions as one of its primary imperatives.

Notes

1. Final Report.
2. Various forms of pressures of this kind were analysed by Professor Nicolay Petrovich Dubinin (whose paper was distributed to the participants, even though he himself was not present at Varna), and by Dr Louis J. Verhoestraete.
3. In industrially developed countries and in the developing countries, as was shown by Dr Louis J. Verhoestraete.
4. Dr Shankar Narayan insisted on this point, which was commented upon at some length in the ensuing discussion. We shall return to it later.
5. The paper presented by Professor Vadim Lvovitch Deglin (distributed to the participants in his absence and subsequently published in the *Unesco Courier*, January 1976) and the contribution by Professor M. E. Vartanian were very enlightening on this point.
6. Final Report.
7. There is still, however, considerable inequality: infantile mortality, which is around 30 per mille in industrially developed countries, may rise to 65, 100 or even 200 per mille in less developed countries.
8. Final Report.
9. In fact, AID is too recent a development for it to be possible to appreciate (on the basis of a large number of cases) its effect on the relationship between the 'adoptive' father and the adolescent child.
10. The Final Report refers to the various precautions to be taken.
11. It entails in particular a redefinition of the notion of 'next of kin') and a thorough examination of the concept of the 'gift of life'.
12. Paper presented by Dr Vasken Der Kaloustian.
13. ibid.
14. Without going into the question in depth any more than was done at Varna, a number of other considerations ought to be mentioned.

15. Final Report.
16. ibid.
17. ibid.
18. Final Report, echoing the contribution by Dr Harmon Smith.
19. In truth, serious biologists are fully aware of this, as was vividly illustrated at Varna.
20. They were at the core of the paper presented by Professor Stéphane Anguelov.
21. Introductory paper by Mr René Habachi.
22. ibid.
23. ibid.
24. Is this not the meaning of the term 'functional ethics' used in the Final Report?
25. This point was forcibly emphasized in the introductory paper and the Final Report.

The order
of living beings

The 'de-termination' of the 'good'

The dynamics of life

There is no such thing as life; there are only living beings. This truism is well worth bearing in mind at a time when artificial means and technical methods tend to oust 'nature'. Life cannot be isolated or conceived in the abstract. Indeed, life is inseparable from matter—it *is* matter—and, in this sense, we may analyse its constituents and processes. These constituents and processes are coded, and we may even succeed in 'cracking' the code. However, we shall never understand how the code originated; and, even if we were able to, the fact would still remain that life is in the process of being transmitted from living being to living being, that it is apparent as the sum, the system of the actions and reactions of these living beings in relation to each other and in relation to the environment. By 'life' we primarily mean a chain, an impulse, a dynamic process.

In this perspective, it is not simply that life will never be reduced to a formula; it can never be apprehended as such, since the very act of perceiving life is itself a manifestation of life. We can only show (not demonstrate) life; we can do no more than indicate its 'mechanisms' and conditions. We cannot even conceive all its potentialities, since they emerge in time, on an everlasting line between past and future (and the stoppage of time would mean the cessation of life, but would not, in theory, imply that all the modalities of life had been deployed). Even if the impossible happened and we one day managed to reconstitute all the necessary physico-chemical conditions so that the resulting actions and reactions could be seen organizing themselves in the shape of a living being, life would still elude us. There are two possible reasons for this: either we would determine such a biosis according to our own purposes, so that it would be no more than an extrapolation of our life, or this biosis would embark upon a history of its own, leaving us merely to observe the ensuing phenomenon. Ultimately, life cannot be comprehended. A comprehension of life would imply, overall, that we lived and existed outside life, and that life could be frozen or set in space and time. Life can only be known through recognition.

The primary fact to be recognized is that life originates in an act of 'creation'. There is not, strictly speaking, creation in terms of simple causality: the cause extends according to its own order into the effect. Something is created when effect and cause pertain to different orders. All biologists (whether materialists or spiritualists) agree that, while the living being is a synthesis of physico-chemical constituents, life and organic matter pertain to different orders: for one thing, the matter would appear to obey the fundamental principle of entropy, whereas life implies negative entropy.[1] This means that we shall never grasp the origin, the cause, of life (whether in terms of 'chance' or 'necessity'). Even if we were to determine what conditions are necessary for the emergence of life, various unknowables of an even more radical kind would persist—between the conditions for the emergence of life and the actual emergence of life, between its emergence and continued development.

If this is the case, we shall always see life as a phenomenon turned in upon its own order, which antecedes its own final cause.

These fundamental considerations[2] give rise to three propositions in particular, essential to our theme, which are all too often neglected.

First, as life is a dynamic process (what goes on within living beings), it is wrong to regard it as pertaining to an intangible 'nature'. Undoubtedly the manifestations of life conform to a profound necessity (otherwise they would not survive nor would they be reproduced); they unfurl according to a 'logic' and are characterized by an overlap of manifold structures and functions. But this 'given' is also a 'giver' which has unfurled in various forms over the millenia. Hence we cannot, without a certain arbitrariness, either advocate an unconditional respect for what is 'given', or purport to break free from the necessities or laws, the code and the logic which remain the foundation of our existence.

In other words, any approach to the question of the 'good' in relation to life presupposes the maintenance of a dialectic tension between the 'given' and the 'giver'. And in so far as we aspire to act on living beings, our task is to modulate 'what is' in terms of 'what may be', and conversely.

Second, there can be little doubt from the preceding paragraphs that life will always evade us.[3] Nevertheless, even if we cannot control life, we can—and must—at least attempt to influence the way it develops, and this quite apart from social considerations or considerations of a personal nature (e.g. the 'freedom of the individual'). It is certainly part of the 'logic' of life that we should guard against the dangers which threaten it and the failures it may engender and seek to improve its quality and the satisfactions it affords. More radically, since billions of constituents combine—partly, it would seem, at random[4]—in even the simplest living being, it is equally part of the 'logic' of life to seek more suitable combinations. Nevertheless, we cannot define an optimal state *a priori*. The reason for this is not simply (we shall have occasion to return to this) that, on pain of degeneration or death, it is essential at all times to preserve singularity and adaptability to changes of surroundings or environment, but also to define this optimum would amount to inhibiting the dynamics of life. By seeking mere improvement, we would forego (for ourselves and for future generations) the possibility of acceding to qualitatively higher forms of existence. Hence, in biology the 'good' is not immediately definable in terms of perfection and any ethical approach must preserve the tension between improvement and regeneration at a qualitatively higher level.

Finally, if we consider living beings globally, it is evident that life is confusing (and confused): at one and the same time random and progressive, appearing to obey both chance and necessity, self-destructive and self-preserving. This is so much so that it seems impossible at the outset to understand the final purpose of life, which only becomes apparent *a posteriori.*[5] Furthermore, progress and evolution depend upon increasingly elaborate systematization or complexification. The fundamental question in relation to life is accordingly 'How?'; and it is precisely this question which is at the heart of scientific research.[6] Scientific research seeks to understand how something 'occurs', how something functions. Undoubtedly, biologists cannot do this without asking the question 'Why?' (or 'Wherefor?'), that is they have to ask about causes. But these questions are subsumed within the question 'How?' and bear upon the sequence of phenomena. Any presupposition is out of place here: the essential is to refine the knowledge of 'What is'.

Applied knowledge is a completely different matter. It seeks to investigate origins (of this or that disease, for example) and pursue an object (a cure, an improvement or an invention); the question 'How?' arises only subsequently. A decision is taken to carry out a transplant, and the feasibility conditions are studied in the light of this decision. In other words, a value judgement informs reality, a final goal is 'pro-pounded'. The question 'Why?' concerns the criterion involved in the definition of life; the question 'How?' relates to ways and means.

These observations may appear somewhat specious. In our view, however, they provide us with a basis for understanding one of the root causes of the uneasiness felt by a good many biologists. Those who are more involved in 'pure' research are worried at the use which is made of their work. Mindful of the system and sequential processes involved in living beings—the 'logic of life'—they call for restraint on the part of scientists, caution and a rigorously critical approach so as to put into perspective certain projects and ventures which should not be implemented until we have adequate knowledge of their implications and medium-term or long-term effects. For their part, scientific practitioners and others engaged in applied research—alive to the dangers and deficiencies which threaten individuals and communities, anxious in consequence to bring about an improvement in living conditions, and under pressure (often also from their employers and the general public) to meet what seem the most urgent needs—are at times forced to rush into experiments, the repercussions of which, despite their irreversibility, are unknown to them. To those who accuse them of rashness they retort, not without reason, that life explores itself; that it incessantly invents new forms of living being, new modes of subsistence; that it devises the antidotes to the risks incurred; that it develops its own defences and secretes the means of its own proliferation. Nevertheless, this continual inventiveness does not proceed without disorders or aberrations; in particular, it requires time, which is so grudgingly meted out to us today. Man would take it upon himself to accelerate processes which, if left to 'nature', would need hundreds or thousands of years.

We do not presume to settle this dispute between opposing tendencies in these pages. Moreover, this dispute is intensified by the fact that the biologists themselves, whether research scientists or practitioners, belong to different cultures and support particular views about man or ideologies. Their ethics are thus differently orientated. Nevertheless, given the power of the resources with which man is endowed, it appears more and more evident that, if it is for man to replace the hazards of existence or evolution with artificial devices, and to facilitate the advent of that which remains potential, he cannot do this without regard for the laws governing life. The necessary point of departure for any ethics presupposes that the 'good' or 'perfection' aimed at should not be devoid of a profound concern to enter into the 'logic' of the living being; that the 'why' or 'wherefor' of our interventions should be modulated according to the 'how' of the phenomenon of life.*

We touch, here, upon a question that is crucial to the whole of this study. Are we not, by laying down as a principle that ethics should enter into the 'logic' of the living being, taking a step which reduces human specificity? Can man be defined solely in biological terms? Does he not emerge according to a completely different 'order', betokened by the original and originating dynamics of his reason, the supreme manifestation of which would be his reason, the supreme manifestation of which would be his freedom? There is evidently, in this respect, a fundamental split between the various philosophies, particularly between the 'materialisms' and the 'spiritualisms'. Nevertheless, regardless of the

* Very close attention will have to be paid to the methodology and philosophy of biology . . . For new knowledge to emerge in regard to the problem of life, it will be necessary to develop a theoretical biology which will reveal the essence of life as a qualitatively unique phenomenon.— *Conclusion of the paper presented by Professor Dubinin.*

deepest convictions in this matter, the fact will always remain that man is his body, a living being, just as life is matter and energy, even if it unfurls according to another 'logic'.

Moreover, even if it is maintained that reason and freedom are essentially different from life, it must be recognized that both exist only in incarnate form.[7] If reason gives understanding, it does so only to a living being, even if it aspires to bring the latter to a higher level of being; and it is always (albeit in general unconsciously) according to a 'logic' which reproduces, even when surpassing, the order of life from which it emerges (although this 'logic' may of course be affected by the vicissitudes of individual physiology). As regards freedom, is it necessary to recall that freedom is by definition relative (freedom in relation to a particular mode of determination) and that if it is to remain true to its nature, it can only be exercised by according value to the very thing from which it frees itself (in other words, remains tied to the body)? Thus, it would appear that freedom has the fundamental obligation to enter into the 'logic' of the living being, to make it ours. In point of time, the 'how' of life takes precedence over the 'wherefor' or, rather, the 'wherefor' or 'why' (even if we are thinking of the 'final purpose') can only be reached through the 'now'.

It should also be recognized that there is no such thing as the 'purely social'. Not that the 'logic' of a society can be accounted for in purely biological terms (just as a knowledge of bacterial structure is insufficient to account for the infinitely more complex structures which constitute the organism of a mammal: there is, in this case, a structuration of structures); in the case of human beings, the complex workings of reason and freedom also come into play. But it cannot be forgotten that man would not exist if each of his cells, however diversified they may be, did not reproduce, in the essentials, the inner functioning of the bacterium; or that society—or, to put it more exactly, the 'social body'—is composed of living beings who tend to organize themselves in order to survive. However specific it may be, the social 'model' inherits from the biological 'model'.[8]

These are primary truths: however, we continually fail to recognize them. Would not the economists be shocked to learn that the manifold forms of exchange which they endeavour to 'regulate' have their archetype within the cell and in the relations between the living being and the environment: life is production and consumption, management and profit, harmony and thrift. In the same way, traditions and cultures have an analogous function in societies to that of the 'memory' of the living being, which is essential to its survival: 'horizontal' memory, enabling it to prepare, execute and co-ordinate its reactions; 'vertical' memory, enabling it to preserve itself, reproduce, adapt, evolve.*

If it is true that there is no such thing as the 'purely social', it is essential to reconcile orders or, if one prefers, 'logics' which we today tend to treat separately: the biological order, the social order, the economic, order, etc., are each investigated in and for themselves increasingly divorced from one another until they become 'delinquent'[9] in relation to each other and do each other violence—even though biology is, as we have seen, more and more integrated with the whole range of economic and social activity. Biology is 'delinquent' if it pursues objectives which are at present inassimilable by the social or economic order; particular instances of this 'delinquency' would be the widespread use of an excessively costly restorative medicine (to the detriment of preventive medicine), and the prolongation of life to the age of 100 or 120. Conversely, the economic order is 'delinquent' when, for example, industrial pollutants multiply. We attach a constantly growing importance to socio-cultural and socio-economic studies; it is no less necessary to develop on a joint basis bio-social, bio-cultural and bio-economic research. Economists, sociologists and political scientists cannot neglect biology insofar as certain practices or ventures weigh upon us, compelling us to delve deep down within ourselves in the search for a new somatic equilibrium. For his part, 'as a member of society, the biologist must redefine his obligations and social functions . . .'.[10] In so doing he must endeavour to take a critical view of the use made of his research, in terms of what is needful for the life of the 'social body', more especially since what is needful in this respect may serve as a pointer to fundamental physiological functions in the individual organism, and must also force those who determine the social superstructures to

* Let us not thoughtlessly cast off our traditions, myths or customs like shop-soiled garments whose individuality was the source of so much rivalry. Let us rather see in their similarities, beyond all culture, the ecologically differentiated outer casing comprised by the various types of behaviour which form the basis of the species. The discernment of this ultimate human reality is essential to the struggle against the behavioural deviations made possible by the abundance and precision of our empirical, experimental and intellectual attainments. Let us guard against false caution which would lead us, in the name of a third-rate scientism, to dismiss the multimillenial intuitions of man: they are still able to afford us protection against the risks inherent in knowledge.—*Conclusion of the paper presented by Jacques Bril.*

respect the vital infrastructures, or rather the 'logic' of the living being.

The test of compatibility

This 'logic' is itself governed by certain primordial conditions which cannot be disregarded without seriously disturbing the order of living beings.

Symbiosis

The first of these conditions is symbiosis with the environment and with other living beings—a subject about which our contemporaries are particularly sensitive. This theme would call for extensive development; as we cannot undertake such treatment here (a special volume would be needed), we shall merely draw attention to a number of problems which arise in this context and which, in our view, are of major importance although the general public is not so familiar with them.[11]

Everyone knows that viability is tested and defined by a combination of actions and reactions with and within a particular environment. By 'particular environment' we mean, more especially:

An environment specially suited to the appearance and development of life in that it contains all the elements necessary to its constitution, subsistence and regeneration (in this connection it is not only physico-chemical constituents which play a part, but also temperature, the presence or absence of radiation, etc.).

An environment transformed by living beings, which cause (or have caused) its composition or decomposition, enriching it with their deposits. Thus it is that, for more than three and a half billion years, the earth's crust and the atmosphere have been untiringly and prodigiously 'worked' by the most elemetary living beings (bacterial or pre-bacterial) to which we are indebted, for example, for the oxygen in the air, the transformation of certain ores and the formation of the oil-fields.

An environment which selects and modifies life, in particular by eliminating the least fit.

An environment in and through which living beings enter into relation (this is a fundamental function since, as the scale of living beings is ascended, life increasingly lives on life and constitutes its necessary environment).

Evidently, these four 'functions' of the environment are interdependent and interactive. More precisely, living beings and their environment form a macroscopic system.

No one today is unaware that if we disturb this system by exhausting this or that of its constituents or by overloading it with certain substances, chain reactions are set in motion which in the long term may prove to be catastrophic. Regeneration of the eco-systems will undoubtedly be one of our major tasks in the coming decades. It is true that it falls to man, as to every living being, to transform his environment materially. But he must understand and preserve its function(s), compensating any imbalance which he may artificially bring about through excessive intervention. It is for having disregarded these functions that our contemporaries are faced with those problems of pollution, and the degeneration of terrestrial and aquatic fauna and flora, which are each day the subject of so many cries of alarm.

Here, our previous observations are borne out with particular force. On the one hand, once he intervenes in life directly or indirectly, man is obliged to modulate the 'why' of his interventions to the 'how' of the system. On the other hand, he must guard against being carried away by the satisfaction of our needs, even if they are completely legitimate: what is initially seen as an improvement should be subjected to criticism and viewed against the global dynamics of life.

In truth, the modification of the environment is carried out incessantly and necessarily in manifold ways. It is about some of the latter that responsible biologists are anxiously asking themselves certain questions. We shall confine ourselves to three of the ways in which the environment is modified.

Is it sufficiently well known that pollution is not simply due to the discharge of toxic waste into the atmosphere, on to the land or into the waters? It is in no way our intention to minimize the threat of eco-logical catastrophe, but this is something of which our contemporaries are aware. On the other hand, they may not be sufficiently warned of the risks which man himself runs from exposure to certain forms of radiation or active chemical substances, such as the alkaline compounds, many weed-killers or pesticides, etc. The human environment is in the process of being saturated with such substances (DDT is even found in the muscular fibres of the wings of penguins in Alaska).[12] In addition to these, there are the drugs which we absorb more directly, such as tobacco and alcohol, and also the various medicines taken to mitigate disease, the abuse of which may seriously jeopardize health (or the health of the embryo if such medicines are taken during pregnancy).* This led Dr Verhoestraete to state the following at Varna:

'It is imperative to assess, from the health point of view, the carcinogenic and embryotoxic effects of these substances and to determine their potential mutagenic action since this would influence the genetic structure of future generations. The results of chemical pollution of air, water and food calls for a system of continuous environmental monitoring. It demands the creation of reliable epidemiological indices for environmentally-induced health hazards, the development of international collaborative studies on man and lower animals and, in general, the strengthening of research in these fields to determine the possible mutagenic, carcinogenic, embryotoxic and other adverse effects of these pollutants.'

For his part, Professor Dubinin concluded his treatment of the various forms of pollution as follows:

'A crucial aspect of this ever-recurring problem is the possibility of damage to man's genetic composition. The task of monitoring the genetic load and the rate of mutations in human populations is a colossal one. Of all the problems connected with the biosphere, this is the one which is destined to occupy a central position in the very near future.'

Another environmental aspect receiving greater and greater attention from biologists is the question of the repercussions of urban life on the psychosomatic balance of the individual. Indeed, man has created in the city an oppressive environment where new threats to this balance evolve imperceptibly yet relentlessly, partly cancelling out the benefits of development, in

* Thanks to the fact that within his social environment man has succeeded in freeing himself from the influence of natural selection, each individual has acquired biochemical and morpho-psychological individuality. The effect of drugs in such circumstances cannot always be predicted. In many individuals genetic changes occur at various links of the metabolic process, and this in itself is not significant. But the absorption of certain drugs may make them ill. The reason for this is that their biological characteristics are drawn into the metabolism of the drug. There has even come into being the concept of 'drug-induced illness'.—*From the paper presented by Professor Dubinin.*

particular better medical treatment and the decline in mortality.[13]

Apart from accidents which affect all ages but especially the younger and productive age groups (with the highest accident rate being found in highly industrialized and urban centres), the majority of deaths may be attributed to three main groups of diseases: cerebro- and cardio-vascular disturbances, chronic respiratory disorders and malignancies. Undoubtedly one reason why these conditions have increased in importance is the greater life expectancy of the population: until recently)[14] when most people died before the age of 30, generally as a result of infectious disease, their arteries did not have the time to sclerose and their organisms (like those of young people today) were less prone to cancerous disorders. We ought not, therefore, to make comparisons involving a dubious factor; nor ought we rashly to blame the industrial and urban environment. Nevertheless, it is certain that manifold factors combine in this environment to contribute to the disorders to which we have called attention.[15] Apart from exposure to various physical, chemical and even biological agents the following factors are significant: a lack of exercise; working and transport conditions which are nervously exhausting, if nothing else; inadequate diet (food additives, dietary deficiencies and overeating); the abuse of drugs and hormones; stress brought about by manifold forms of aggression suffered by the inhabitants of the big cities, etc. It seems that the combination of these factors (which also affect the embryo) is responsible for an underlying genetic susceptibility. The latter may, of course, have other causes linked, in particular, to excessive nervous fatigue (especially in women), behavioural problems and promiscuity: it is known that animal species decline when overcrowding occurs, and that colonies of insects split up after excessive proliferation. Such observations led Dr Verhoestraete to declare at Varna:

'The possible influence of present living patterns of urbanized societies on resulting psycho-social problems makes it necessary to intensify genetic, biological and socio-cultural research on as broad a base as possible. Corroboration of the increase in psycho-social problems and improved knowledge about their causation and prevention might strongly influence concepts and thinking about the optimal organization of communities in the developing countries undergoing industrialization and lead to reorganization of the existing living patterns of urban settlements in presently industrialized countries.'

Other factors (to some of which no more than passing reference was made at Varna) accentuate man's overall dependence on his environment and the need for him to transform it. Furthermore, such transformations must not involve either any disturbance of the biosphere or any serious organic imbalance. It would accordingly seem essential to elaborate a genuinely international policy in this field through interdisciplinary consultation. But are those responsible for the various national economies, and are the biologists themselves, ready to come into line with such a policy? We should not be blamed for having our doubts on this question. Today, mankind is in the process of assuming unparalleled powers, which will inevitably have considerable repercussions on the environment and on living organisms: we are referring to the new techniques of genetic manipulation.

What does this mean? Let us consider the question schematically in three steps.

The information necessary to the functioning and reproduction of each cell is coded in the genes. These are distributed, as pearls on a necklace, along the incredibly compressed strand which forms the DNA molecule. Although this code is fundamentally identical for all living organisms, the specific evolution of each living organism has resulted in a barrier preventing any exchange of DNA (and, consequently, of genes) between cells belonging to organisms of different species. In recent years, biologists have been able to cross this barrier: they have not simply succeeded in isolating the DNA molecule within the cell, but have also separated the DNA strands and reconstituted one of them by introducing and 'welding' into it fragments of another strand taken from a cell of a different species.[16] They thus obtain a hybrid DNA molecule, known as a 'recombining molecule'.

If the choice is made to 'recombine' in this way a DNA molecule which will be capable of multiplying more or less autonomously in cells where it is present, a 'vector' is obtained which can be introduced into the receptor cells.[17] this is known as 'transduction'. In this way the recombining molecules will be able to multiply.

Three possibilities are accordingly available, either: (a) to aim at a simple process of amplification (once the cells have multiplied the recombining molecule that has been introduced, it is recovered to the extent that it possesses useful properties—this process is contemplated in particular, for the production of certain vaccines); or (b) to intensify certain specific qualities within the cells or, inversely, endow them with required properties; or (c) to modify the genetic inheritance more broadly, by bringing about desired mutations; this may also be done with certain higher forms of organism.

A considerable field is thus opened up by the possibility of introducing new genetic information into the living being. In the foreseeable future, these techniques will be used in the manufacture of insulin (essential in the treatment of diabetes) and other hormones; for the production of new vaccines which are more reliable, less costly and aimed at viruses that are still unconquered; and to replace chemical fertilizers with nitrogen-fixing bacteria. Looking further, we can anticipate their use for the improvement of certain plant species and, more problematically, the 'correction' of genetic disorders'[18] These techinques have also led to an increase in research aimed at creating bacteria capable of dissolving certain products, notably petroleum (the cleaning of tankers) and various pollutants (the recycling of waste water), or processing other products so as to yield silk (which derives from an enzyme), sulphates (from sulphides), methane and even light.

Research is currently beginning or proceeding in these directions. However, a number of serious biologists are coming to think that, in the long term, these new techniques will make it possible to restore the environment, with man directing or channelling the immense labour of the micro-organisms in the air, on the ground and in the soil. To put bacteria at the service of mankind would be a revolution of incalculable implications (to grasp its significance, it is, perhaps, necessary to recall that bacteria experience binary fission approximately every thirty minutes; if none of them were to die, there would be a million bacteria within ten hours, and in fifteen hours there would be a population of a billion.

Apart from its scientific importance, this kind of research galvanizes powerful interest, accentuated by international competition: funds are generally forthcoming from official bodies and private industry who are anxious to make use of such discoveries.[19] In view of this, certain scientists believe that too little attention is given to the dangers of an accidental rapid dissemination of bacteria carrying hybrid molecules, which could affect man by bringing about aberrant genetic information, or an infection resistant to antibiotics; or which could devastate flora and fauna, and even cause irreversible imbalance in the biosphere. These same scientists regret the fact that these manipulations are carried out with such extreme haste, with inadequate precautions being taken in connection with installations (special premises), personnel (education and training) and the scientific aspects 'neutralization' of the combinations or vectors).[20]

The truth is that the risks involved are inadequately defined (and partly unforeseeable). There are two kinds of danger: the possible escape of bacteria or a pathogenic virus against which there would be no defence; and the disturbance of certain systems. Although the first danger cannot be ruled out in theory, it can be guarded against by various measures involving physical containment (as applied by large institutes and laboratories where high-risk microbes and viruses are manipulated) and biological containment (by only working on 'hosts' which cannot be pathogenic in relation to man), if the choice of projects is properly controlled by specialized bodies. To put certain alarmists' information spread among the general public into perspective, let us recall that no bacteria can withstand a good washing with disinfectant or exposure to ultra-violet rays. On the other hand, it is extremely difficult to evaluate the long-term systemic effects of this kind of manipulation. If oil tankers were to be 'cleaned' with bacteria, what would happen when the resulting biomass settled on the plankton? If nitrogen fixation is used with certain plants, considerable nitrification will result. Will denitrifying bacteria then be effective enough to restore the equilibrium of the nitrogen cycle?

In reply to the appeal made in July 1974 by the Group of eleven scientists led by Paul Berg for work in a number of areas of molecular genetics to be discontinued, the Asilomar Conference in February 1975 adopted a realistic approach. The eighty-five participants confined themselves to calling the attention of specialists to the need to observe precautions aimed at ensuring maximum safety. In fact, the potential benefit of genetic manipulation is such that mankind cannot be deprived of it. But only calculated risks should be taken. Is this not what is done in other areas where potentially lethal techniques are used? This conference cannot therefore be blamed for not prohibiting genetic manipulation, and it must also be given credit for urging scientists to scrutinize their working methods (the International Institute of Health in the United States subsequently issued the appropriate rules in December 1975). Nevertheless, the following questions may be asked: Have the various bodies been sufficiently critical of current research and the opportunities made possible by this biological 'revolution' from the point of view of their social implications? Have they not merely endorsed scientific rivalry and competition, without also calling for co-operation? And have they not consequently given too much of a free hand to unilateral interests (corporations, groups or nations)?

We thus come back to the need to draw up a genuinely international policy in this field, underpinned by a relevant ethics and professional code of conduct. Is this not what the committee of scientists led by Paul Berg was calling for in 1974?[21] However, its representations seem also to have been inspired by another concern which we wish to echo, as it raises a major problem related to our theme. For the first time in history, scientists, conscious of their own powers and the potential scope of the good or harm they can do, have offered to hold back and take the time needed to clarify their objectives and the means to achieve them. They wish to develop the techniques and standards which would make it possible for their research to be carried out in the optimum conditions and directions. 'To take Time' . . . The temporal dimension would appear to be one of the basic elements in the relationship between biology and politics (or ethics). We should at least sketch out certain aspects of this dimension.

Synchrony

Upon careful reflections, we see that one of the most
disconcerting traits of life is its continuity. In effect,
there is every indication that this trait is so unusual in
relation to matter that living beings emerged originally
in a very small number of forms, from which all existing
living beings have descended. However, it is known
that organic systems conform to a rigorous internal
necessity; duration is a constituent of this necessity
(translated in terms of the species by longevity).
Elementary forms of life live for a very short time:
bacteria and certain varieties of cell in the human body
live for about thirty minutes, whereas other cells live two
or three times longer, and nerve-cells (which do not
reproduce) live in most cases until the individual dies.
How did the longevity of living beings increase?
Above all, how could living beings with dissimilar life
spans unite to form complex organisms? We can
imagine that beings coupled or united in space develop
exchanges whereby they are mutually transformed.
However, the fact that these living beings come to form
a body presupposes more than a simple coming-
together. It presupposes a functional companionship,
following a secret long betrothal; an unhurried inter-
course in the course of time. While saying this, we
should not forget that each living being preserves its
own internal necessity (otherwise it would have died).
The appearance of complex new forms of life marks an
astonishing victory over the heterogeneity of the
specific life spans and continuances of each living
particle integrated in a new synchrony. Thus it is that
there is irreversibility. Thus it is also that temporal
succession can occur in the case of each living being in
evolution.

Time was also necessary for living beings to learn to
protect themselves against all the various dangers which
they faced from without (physical and chemical agents,
other organisms) and from within. Time was needed
for the formation of this inconceivable bundle of actions
and reactions, activations and inhibitions, and manifold
forms of regulation which are prerequisites of survival
(it has been calculated that, in the half hour of their
existence preceding binary fission, bacteria have to
perform some 2,000 chemical reactions, most of them

sequential). Admittedly, natural selection has preserved those living beings which have the greatest chance of defending themselves (perhaps fortuitously). It is always the case that adaptation is the fruit of obscure metamorphoses which feel their way and occur in each living being and only prove to be fully effective after several generations? And what about the gradual formation of organs such as the eye, which requires a synchronous transformation of the entire organism?

A biological treatise would call attention to other aspects of the life–time relationship. We shall confine ourselves to these brief observations. Is it necessary to remind ourselves that they apply to mankind? Unfortunately it is, for we continually fail to appreciate this temporal dimension. Thus, unconsciously (for example, when polluting the environment) or consciously (in drug abuse, ingeniously forcing our way through the protective barriers of our various cells), we compel our organisms to absorb more and more substances against which they have not learnt to defend themselves, and even substances against which they have always defended themselves, with the aim of achieving a particular effect (such as soothing pain), without any knowledge of the effects which such absorption might have, in the foreseeable future, on the individual or his descendants. We modify this or that function, without even suspecting the chain reactions which we may unleash in our organism and the 'feats' which we compel it to perform to restore homeostasis simultaneously.[22] We rely on the organism to compensate in areas where we have been unable to make any provision; we expect that it will be restored or re-established identically, whereas we have brought about something irreversible.

These observations also apply to the 'social body'. As a result of the general application of certain advances in contemporary biology, society has to take serious steps to restore its equilibrium. We have given several examples in this connection: the population explosion resulting from the decline in mortality, and the foreseeable consequences of an increase in the use of artificial insemination with donor sperm.[23] To take another example, advances in gerontology may make it possible for human life to be prolonged to the 100–120 year range.[24] Is any assessment made of the consequences of the premature general application of this research? Even if this prolongation did not impair mental faculties, could it be integrated into the overall socio-economic and socio-cultural framework? Persons over the age of 60 already account for 15–20 per cent of the population in developed countries.[25] This modification of the age pyramid (life expectancy was only 31 years in Europe at the beginning of the century) imposes on societies strains and stresses of a unique kind: tension between generations, a delay in the assumption of responsibility, the increasing economic burden of supporting services for the elderly, the dumping of old people, etc. Any further increase in longevity would necessitate a restructuring of the entire economy, and a complete re-thinking of our education systems (with more emphasis on personal creativity and free activities), our labour legislation, and even the family (problems of authority, inheritance, estates, etc.).

This example highlights the importance of the duty to forecast effects,[26] but also the very considerable difficulty which such a task involves. Generally our forecasts are simplistic; more often than not they are restricted to mere linear progression (even in our most elaborate attempts at simulation). They extrapolate quantitatively and qualitatively, from what we know, what we intend to achieve or think should occur; they aim at a global effect. But everything connected with life (organic and social) is rigorously systematic and involves a conjugation of durations. However, not even in society do the various constituents necessarily evolve at the same speed and in the same time. And the delay or advance of this or that part of the whole means that the balance of the whole itself is affected.

Does this mean that we are unable to predict the impact of our undertakings, and are condemned to do nothing? Evidently not, for any tendency to stand still is sclerosing, if not fatal. We should incessantly adapt and protect ourselves. We should guard against the disorders generated by our artificial techniques and products. But it is essential to be fully aware that all progress presupposes a certain imbalance, which must subsequently be compensated, according to an irreversible process. Science does not escape this rule.* We are at least entitled to wait for scientists, particularly bio-

* Science develops unevenly. There is always one branch or other of science which not only is in advance of the others but also paves the way for a leap forward in other branches of human knowledge, not only changing man's ideas about himself and his fellow beings, but also having a significant impact upon society . . . —*Extract from the paper presented by Professor Frolkis and Professor Bezrukov.*

logists, to step back and put their concept of progress
into a temporal perspective, since what is good in itself
may prove to be catastrophic if the time factor is
disregarded. Moreover, any distortion in terms of time
(whether it is a matter of unwarranted continuance or
untimely innovation) gives rise to serious biological
and social disorders. In any case, premature general use
or precipitancy in the application of this or that new
technique should be weighed rigorously in the balance.
In regard to life, precipitancy is not merely a source of
error, monstrosity and maladjustment. It sets in motion
an accelerating process of compensation and over-
compensation so that, in the end, synchrony is no
longer possible; nor, consequently, is life.

Ethical demands

The preceding observations and examples may appear
to have little in common with our theme. What do they
mean in ethical terms? They imply a certain number of
fundamental propositions.

If it is true that there is no abstract life, that life is an
evolutive dynamic process, there cannot be a theoretical
good or a pre-established law which would enable a
purpose to be assigned to biology or biological research
to be criticized on an *a priori* basis.

It would be more accurate to maintain that the task
of the moralist, under the circumstances, is to put
biologists on their guard against the temptation to
pursue excessively restricted aims, imposing on life a
'wherefor' without any concern about the 'how' and
also against the premature determination of a 'better'
or 'best' which would deprive life of its plasticity and
the possibility of passing to qualitatively higher forms
of existence.

In other words, our opinion is that ethics should
aspire to 'de-termine' life rather than define it.

In this perspective, ethics imposes upon us the duty
to 'de-limit' biological research and its practical
applications, notably in relation to the demands of
symbiosis and synchrony; and also a duty to conjoin
and to integrate, after the fashion of the profound
dynamism of life which tends to reconcile, instead of
setting against each other, the parameters discussed in

the previous chapter which may appear to be mutually opposed: survival becomes manifest and materializes in an incessant toil aimed at conjoining the living being and the environment, parts and the whole/collective and individual/adaptation, renewal and continuance/ freedom and constraint/conservation and reproduction/ present and future. There is no mysterious design in all that: it is a fact, an order of life, and as such binding on anyone who sets out to assume control of life. But how is such conjunction to be achieved?

Throughout evolution, an incessant search for compatibility has been conducted among living beings and species, in symbiosis and synchrony. Compatibility has been put to the test at length, slowly, not without errors leading to degeneration or death. The question 'How?' has put the question 'Why?' into perspective; what is useful has been modelled on what is suitable. Biological research pursues this search; the application of its discoveries could not abstract from it, as it is only in this perspective that this research can have any meaning, improving that which exists or leading to a higher state.

Can one test compatibility with any certainty when, as we have seen, any accurate forecast is impossible? It is evident that biology involves an element of risk, in so far as it is not confined to the knowledge of what exists but purports to intervene with regard to the living being.[27] Life constantly involves risks and evolution is an amazing adventure. Ethics does not take away the risks or the spirit of adventure. Although it gives the scientist a duty to respect the demands inherent in the profound necessities and the continuance of living beings, far from enclosing him in the framework of this 'given', it urges him to understand it as 'giver' and to facilitate the advent of qualitatively higher forms of existence.

The fact remains that the latter are not predeterminable (and should not be predetermined): they presuppose that we go beyond ourselves. Accordingly, the ethical approach to the tasks and purposes of biology consists fundamentally in our view, in urging scientists to discern and continue in the direction in which life has progressed, in other words to promote the further development of its 'logic'.

In the following chapters, we shall endeavour to

identify, if we can, the direction in which life has progressed and its 'logic' developed.

Notes

1. This law of matter in its entirety is known as the principle of entropy, according to which the energy inherent in this system undergoes degradation, or passes from order to disorder.
2. We shall have to explain them further, particularly when we return to the problem of finality.
3. We shall return to this theme in the last chapter.
4. It is known that the singularity of the individual, even if it conforms to a code and strict necessities, occurs through the 'choice' of several thousands (or millions) of 'markers', from among the billions of those which are possible, and through their combination in an original manner.
5. We shall have to return to this assertion and qualify it.
6. If there is such a thing as 'pure' research.
7. We shall have to return to this question.
8. Besides, more and more sociologists are paying renewed attention to biological research and discoveries.
9. We should recall that etymologically 'delinquency' means the breaking of a link.
10. Introductory report by Mr Harrison.
11. They were presented as such at Varna.
12. This was one of the themes developed in the paper presented by Professor Dubinin.
13. This question was referred to by Dr Verhoerstraete, from whom we have borrowed a number of the following observations.
14. This is still the case in many developing countries.
15. The comparative analysis of mortality and life expectancy in urban sectors and rural areas in developed countries testifies to this phenomenon. Apart from certain regions where alcoholism is still rampant, life expectancy is notably lower in the large urban centres, despite the fact that their inhabitants are more medically informed and more highly medicalized.
16. This separation may be achieved either by the use of mechanical processes and the subjection of the resulting fragments to the action of certain enzymes; or by the exposure of DNA molecules to the action of an enzyme which recognizes and cuts our particular sequences. The 'welding' is achieved by the use of another enzyme (ligase).
17. The vectors used are plasmids or the bacteriophage L; animal or plant viruses could also be used.
18. We shall return to this question in the following chapter.
19. Not to mention the rivalry among research scientists, seeking fame and promotion.
20. At Varna, very similar considerations were expressed in the paper presented by Professor Dubinin.
21. At Varna, the paper presented by Professor Dubinin echoes their appeal.
22. A synonym for 'organic equilibrium'.
23. cf. above, pages 26–28.
24. We shall return to this question in the chapter entitled 'Individuality and Adaptability', echoing the paper presented by Professor Frolkis and Professor Bezrukov to which we refer here.
25. Within a short time, comparable rates will be recorded in the

developing countries, due, above all, to the improvement in sanitation, nutrition and immunization.

26. Already referred to in the previous chapter.
27. We shall return to this theme in the chapter entitled 'Reproduction and Sexualization'.

Organicity and control

The common good

In the view of many biologists, it may have seemed unscientific to present the ethical approach to the problems which they have to face as a 'de-termination' of life, or even as an investigation of its meaning. Does not life blindly conform to the demands inherent in an internal necessity from which freedom can only be attained through death?

Self-preservation is undoubtedly one of the fundamental traits of life. It is, however, necessary to agree on what this term comprises. For the atom is also an integrated system; the assembly of its component particles continues in existence through the interplay of forces, actions and reactions, attraction and repulsion, even if these particles are heterogeneous. And this combination does not have a specific duration: it lasts as long as the relationship of forces continues (except where there is fission brought about accidentally or artificially) and remains passively subject to the law of energy decay. Furthermore, in such a system, each of the component parts may be modified by the action of the others and the whole, but this modification does not involve any innovation. What occurs in the case of the living being is something completely different. The living being is not a mere assembly; in conveying an idea of its unity, the notion of 'system' seems less pertinent than that of 'organicity'. Organicity is characterized by the release of an 'adjunct' of energy; by the 'appropriation' (assumption) of a temporal mode, a certain duration, or rather by the subsumption of the specific durations of each part in the specific duration of the whole (there is not only synchronization, as we have noted, but a definite longevity and distinction between ages); and by a continuous 'invention' of functions and new forms of control (where once there were scales there are now eyes, and so on). Organicity involves an extraordinary dynamic process in each part in the service of the whole and vice versa. Thus, for every organism it can be said that the role assigned to each of its component parts is strictly determined, yet there is at the same time a de-termination or de-particularization of the activity of each of the parts, which are controlled in terms of the whole.

These theoretical questions are important. On the one hand, they suggest that the practical applications of biology should have regard to organicity as well as compatibility. To give a general example, let us refer to the 'aberration' constituted by the mastodons of prehistory. Their fantastic development was compatible, since certain species actually lived for thousands of years. But, in the course of time, this gigantism proved to be inorganic, as these animals did not develop a cerebro-spinal system affording an adequate innervation mechanism.[1] On the other hand, it is manifest, if not that the 'common good' takes priority over the good of the individual parts (it is chasing a will-o'-the-wisp to try to determine which comes first, in order of time or precedence), this 'common good' at least has certain irreducible demands. This observation may appear no more than a truism, in the case of an organism such as a plant or animal. The matter becomes more complicated when this plant or animal is seen in the context of the species. The biologist is faced with various thorny questions as soon as he considers the individual in his relationship with mankind at large, or as a member of the social body. Even though society is composed of an assembly of living beings who adopt irrevocably a certain pattern of organization in order to survive, even though it is incessantly in a state of necessary regeneration (as existing individuals cannot dispense with future individuals, and vice versa, the question remains to what extent it can allow individuals to put their physical resources to whatever use they judge fit or, conversely, to what extent it could constrain them in biological terms. Three major problems inevitably arise in this context: genetic intervention, birth control and the safeguarding of world health.

The genetic endowment of mankind and genetic intervention

Genes control the structure of the organism; they are also responsible for the vital functions of assimilation, growth and regulation. They carry and transmit the code and 'programme' of the organicity of the living being. The chromosome chain which assembles all the genes is transmitted through the ages from generation to generation, in varying degrees of elongation or complexity, carrying the appropriate information for the species. Accordingly, the genetic make-up of man constitutes the fundamental, essential endowment of mankind, the mainspring of its continuance and reproduction.

Gene therapy

It is true that this hereditary endowment is not perfect, either in terms of indivdiuals or groups of individuals, or in terms of the species. Certain human beings have genetic abnormalities which affect all or part of the genes and give rise to serious disorders. These abnormalities or deficiencies are mostly hereditary and, if reproduction is still possible, they may affect entire families, groups of individuals and even an ethnic group or race.[2] This explains the importance of gene therapy.[3] The latter may be performed in two ways.

The first method consists in correcting or compensating the effects of genetic dysfunction. Generations of physicians have employed it; their efforts deserve unqualified support.*

Sometimes, however, these effects cannot be compensated. Or, again, the deficiency, scarcely perceptible to the individual, becomes much more serious, or even disastrous, for his descendants (once more assuming reproduction is possible). This is especially the case when both parents carry the same disease (homozygotes). If they wish to bring a normal healthy child into the world, should not everything be done to help them? Lastly, it cannot be forgotten that the care of persons suffering from genetic diseases often places a very heavy burden on their families and society, especially since this care generally has to be permanent. These are the main reasons behind the tendency on the

* Conditions such as plenylketonuria, diabetes mellitus, severe combined immune deficiency, can all be successfully treated allowing patients to survive and procreate. The conservation of their deleterious genes in the genetic pool of the population has mild dysgenic effects of little significance as against the immediate benefits to be obtained for those affected.—*From the Final Report.*

* This present stock of knowledge of molecular biology opens up possibilities for manipulations of genetic materials that would have been inconceivable only a few years ago. Shapiro, Beckwith and their colleagues have reported the isolation of a set of bacterial genes, the so-called 'lactose' operon, taking advantage of the fact that these genes can become incorporated into bacteriophages in such ways that artificial hybrid DNA molecules can be prepared containing only the 'lactose' genes in double-stranded form. Similar methods can be devised to isolate other sets of bacterial genes.

Another stupendous achievement is the complete in-vitro synthesis by Khorana and his co-workers, of the DNA sequence corresponding to a transfer RNA molecule (the alanine tRNA from yeast) by methods applicable to other genes of known sequence . . .

These discoveries point to the possibility of preparing pure genes or sets of genes in substantial amounts and, coupled with advances in virology, in embryology, and in reproductive biology, may herald a forthcoming time when direct gene manipulation may be possible, not only for bacteria and viruses, but in plants and animals, and even in man—the possibility of genetic engineering.—*From the paper presented by Dr Narayan.*

† But since it is essential that gene therapy be performed without

part of biologists to treat, where possible, the causes as well as the effects; in other words, to practise direct genetic intervention.

Genetic intervention is now a practical possibility. It can take two forms, as we mentioned in the previous chapter in connection with genetic manipulation. First, it is possible to 'crack' the DNA molecule, or to cut a chromosome chain and reconstitute it after transforming one of its 'segments'. There is nothing to prevent the same procedure being carried out with the gametes. Taking a male and female germ-cell, it would be possible to recombine their chromosomes after 'replacement' of the defective gene or genes. This would be followed by artificial fertilization and the implantation of the 'cured' zygote in the uterus. In this way, a normal child could be born. The second form of genetic intervention would consist in obtaining, through reversible transduction, a gene whose messenger RNA could be isolated. 'The introduction of such genes into cells by viral transduction may then be possible. Inter-specific cell hybridization experiments have shown that widely heterologous genetic material of diverse origin can function in many different cellular environments. Thus, genetic treatment of a human single-gene bio-chemical defect could become possible in the not too distant future.'[4]*

As we have seen, present attempts at gene therapy involve a single gene. Will it one day be possible to change several genes? We already know more than a hundred diseases of a serious nature which result from a single defective gene. 'Attempts at gene therapy for specific diseases . . . thus deserve to be encouraged.'[5] But what about the side-effects of such therapy? A gene often carries several characteristics: could the variation of a single characteristic not have serious consequences for the others and, consequently, for the organism? It should also be noted that certain diseases or heredit-ary afflictions are functional: for instance, as sickle-cell anaemia enhances resistance to malaria. Accordingly, the greatest caution has to be exercised.† We have to admit the extremely limited and fragmentary nature of our knowledge in the field of genetics. Moreover, in view of the microscopic proportions of the genome and the inconceivable complexity of the operations and reactions of each of the millions of genes, we may ask ourselves whether it will one day be possible to understand fully the nature and action of the genetic system. We cannot, however, give up our attempts in this direction, if only to combat certain diseases such as cancer.

Towards eugenics?

Apart from the treatment of certain diseases, genetic manipulation has opened up the possibility of direct intervention in regard to the fundamental characteristics of species of living beings. In the case of agriculture, the implications are enormous in view of the increase in the world population.[6] Applied to man, these interventions open the door to practices aimed at the improvement of the human species—either negatively, by the mitigation of certain organic disorders, or positively, by developing various abilities. This advance would be the cause for unqualified congratulations if the same techniques could not also be used for purposes of 'eugenics', manipulation aimed at producing individuals or groups of individuals with specific traits or aptitudes. A good number of scientists make no secret of their concern. In his paper at Varna, Professor Dubinin stated the following: 'The progress which will be made in genetic engineering in the next few decades will create a new situation in biology. It will be a time of untold power over the organic world, giving man the ability to alter his own biological nature by means of carefully planned intervention.'

Most scientists admittedly still consider eugenics to be within the realm of science fiction, because of the complexity of the genetic system and the nervous system (particularly the cerebral system), which both control psychic exprience. With our present knowledge, the creation of a 'superman' is no more than a dream.*

Nevertheless, even if they rule out the possibility of eugenics, a good number of eminent biologists continue to refer to it, if only to condemn it. Thus Professor Dubinin states: 'An axiological approach to man as an exceptional historic and natural phenomenon requires in the first place the identification of his genetic nature and in the second a decision as to whether it can and should be changed. This decision will be made against the background of our knowledge that with his present

causing serious deleterious side-effects in the treated individual or his progeny, extensive work with tissue cultures and experimental animal models will be necessary to evaluate in depth the different aspects and potential side-effects of gene therapy.—*From the Final Report.*

* The eminent molecular geneticist, Jacques Monod, has written in this connection: 'Not only does modern molecular genetics give us no means whatsoever for acting upon the ancestral heritage in order to improve it with new features—to create a genetic "superman"—but it reveals the vanity of any such hope: the genome's microscopic proportions today and probably for ever rule out manipulation of this sort' (*Chance and Necessity*, New York, 1972, p. 164). Monod believes that a race of supermen could only be produced by means of selection, using the methods of classical eugenics. He continues: 'Science fiction's chimerical schemes set aside, the only means for "improving" the human species would be to introduce a deliberate and severe selection.' Here Monod himself seems to have fallen prey to the chimerical schemes of science fiction, as genetic selection for the purpose of restructuring human characteristics—primarily in the 'super-biological' sphere of man's existence—is no more promising an approach than genetic engineering.

Does not such a statement reveal a lack of faith in man's powers of reason, his infinite capacity for knowledge and the transformation of nature? No, this is simply to recognize the relative character of our knowledge of the physical world at this stage in time. We still do not understand the basic laws governing the microcosm and the cosmos, this despite the theory of relativity and quantum mechanics. It has not occurred to us, for example, to try to alter the earth's orbit, bringing it closer to the sun and thereby ensuring an increased supply of energy on the earth's surface; to give the moon soil and an atmosphere; to change the planets' gravitational fields, etc. Man's genetic make-up and the way in which it is achieved through the processes conferring individuality, the creation of his 'superbiological' spiritual faculties—all these represent the highest level of motion of terrestrial matter.—*From the paper presented by Professor Dubinin.*

genetic makeup man has historically shown himself capable of limitless extension of his knowledge. Efforts to apply genetic engineering to the genetic adaptation of human beings are made by scientists who have no sense of professional or social responsibility.'

Such a prospect is fascinating despite the fact that it is a cause of concern and the subject of censure.[7] This is undoubtedly due, in part, to the fact that much of the research in this area is being done by scientists working in agriculture. Experience has shown that the techniques developed in the various fields of agriculture have every chance of being applied to mankind. They would at first be used for commendable purposes such as the treatment of disease, but their use could be subsequently extended to more questionable aspirations at an individual and at a collective level (the typical example here is artificial insemination). Hence, the Final Report of the Varna meeting could not disregard at least the possibility of eugenic practices, or at any rate practices aimed at the improvement of the species: 'However, if future developments of biological techniques enable scientists to attain this possibility, the "manufacture" of man according to specifications would present difficult problems of health, choice and ethics. The problems of choice of attributes may be unsolvable, since the value systems are multiple and the so-called advantages of a specific trait under certain circumstances present disadvantages under others. Furthermore, as the genetic endowment of individuals and populations is the product of a long evolution, it is biologically unwise to alter their genetic structure suddenly and significantly without knowing, with much more precision than is the case at present, what causes have led to these structures and the possible consequences of major modifications.'

As may be seen, it is not merely a matter of inadequate scientific knowledge; various other questions arise in this connection. Certain of them concern the individual: the organic compatibility of these interventions, in relation to the historic character of the organism and personality, adaptability, and integration in the social environment (we shall return to them in the next chapter). Other questions directly concern society: How do we define the functions and demands of society, and what authority does it have over individuals?

E

Society and control

Even if it has its own laws and, in the last anlaysis,
cannot be reduced to a mere biological 'model', society
is, as we have seen, to be regarded as participating in
depth in the dynamic process of life. Society is not
simply a collection of living beings in juxtaposition;
it is the sum of all the relations, interactions and
communications between human beings who organize
their existence in order to survive. In view of this, its
primary function, its basic 'programme' is, naturally,
to ensure the permanency and reproduction of the living
beings which constitute it. It is not that society can
produce a uniform type of individual: the cells and
organs which make up living beings have their own
laws of preservation and reproduction, which may not
be transgressed without bringing about the destruction
of these living beings. Thus individuals and groups must
preserve their own original traits and freedom within
society. Conversely, however, society can only continue
in existence by controlling the way in which individuals,
and mankind at large, develop, biologically speaking.
'Socializing' life does not mean standardizing it; it is a
matter of organizing life, or rather, making it organic.

We cannot escape the necessity of designing a
genuine politics of life coupled with an ethics of life.
However, it would appear that such a politics and
ethics would have to evolve on three levels. These
levels would relate to the individual good (for instance,
the struggle against disease); the good of the greater
number (for instance, the development of preventive
medicine as against high-cost, sophisticated medical
treatment); and the good of society (for instance, the
mitigation of aggressivity factors, insofar as they
disturb the whole skein of social relations). These three
'goods' sometimes stand in direct opposition to each
other. Any politics and ethics of life inevitably involves
the making of choices or the imposition of compromises
with a view to the reconciliation of these 'goods'.
Furthermore, this necessity will become increasingly
acute.

The key concept in this context (as fundamental as
the notion of compatibility touched on in the previous
chapter) is that of control. This is a paramount concept
in biology, since true organicity is only to be found in a

concourse of parts and organs, each surviving and reproducing according to its own laws, each 'functioning' to its own ends, yet controlled by its ordering totality. A typical example is the multiplication or growth of the cells of certain tissues. Throughout the genesis and the early stages of the living being, this growth and multiplication occur rapidly; in adulthood the process slows down if it does not halt (this depends on the tissues and the lifetime of each type of cell). On the other hand, if there is a bruise or a wound, multiplication accelerates so that a scar can form. In the human organism, for instance, such forms of control are innumerable, increasingly rigorous and more and more strictly programmed, as one moves inwards from the periphery to the centre of the organism, from the accidental to the essential.

By the same token, it is obvious that society needs a system of control. Until recently these forms of control remained relatively 'external', however burdensome they may have appeared to individuals (and 'nature' exercised constraints, in numerous fields, which dispensed men from self-determination). Today, nobody escapes these forms of control. At its present stage of economic development (with its incredible technological potential) in an era of extreme urban concentration and with the evolution of increasingly complex relations of interdependence between all countries of the globe, mankind has necessarily to form a single body, whether it wills it or no. This process is irreversible. Over the last thousand years, individuals and communities have structured themselves into nations (and this 'task' has still to be completed in certain countries). The coming centuries will see an intensification of this process, and these nations will become organs of mankind. As a result of this new demands will arise. In the economic sphere, there will be splits, disturbances and imbalance. We shall have to learn that phenomena seen by us as crises are aspects of the search for balance, systematization, and a new world order. There is nothing catastrophic in that. But these demands inevitably have increasingly serious repercussions on the whole spectrum of socio-cultural relations and everything that concerns life: for example, population health, the decline in mortality or the prolongation of human existence, and, above all, the birth rate.

The population problem

We have received no lack of warnings on this score. Today this is undoubtedly the most important problem concerning control, even if only one of a number calling for a radical change of mentality. Mankind cannot continue to reproduce at the rate it has done over the past fifty years. Without pretending to treat the problem exhaustively (that would need a volume of its own),[8] it seems desirable to go into it in somewhat greater detail. In our view it highlights, and illustrates the very notion of control—and, consequently, what is entailed by respect for the common good.[9]

The figures are well known. Professor Dubinin recalled them in his paper: 'In the space of 4,000 years, from 10,000 to 6,000 years ago, the population increased by a factor of almost 18, reaching the total of 86.5 million. Since then, the rate of demographic expansion has increased very rapidly, from the one millenium to the next, from one century to the next, even in the nineteenth and twentieth centuries from one decade to the next . . . It has been estimated that the world's population will double in 30 years and in the year 2000 the figure of 7,000 million will be reached. And what after that? It has been calculated that if the population increases at an average rate of 2 % per annum, as it has done in our century, then in 700 years time there will be only one square metre of the earth's surface per person. Such a rate of reproduction is obviously absurd. By the year 2020 we must reckon on a population of 10,000 to 12,000 million, after which it is clear that further growth of the world's population would be unreasonable.'

There are doubtless areas of disagreement in the various forecasts: the experts differ on how long it would take for the world population to double. It is actually possible that we may see a generalization of the sharp drop in the birth rate experienced in certain industrialized countries (including the German Democratic Republic and the Federal Republic of Germany where the birth rate is below the death rate). However, a steep decline in the population growth rate would be just as disastrous as an unbridled population explosion, especially in regions with a very high proportion of persons under the age of 20: over 60 per cent in most

Asian and Latin American countries. Be that as it may, we must expect the world population to double in the near future: some predict forty and not thirty years, others announce that it will happen in twenty-five years. There is every indication that by the year 2000 there will be 7,000 million people on the earth. Will it be possible to prevent the world pouplation from doubling again in the following fifty years? In this matter mankind as a whole has a serious political and moral responsibility.

First and foremost, mankind needs food. At the present level of agricultural production, if every inhabitant of the globe ate as much as the average Westerner, the earth could feed no more than 1,500 million. A number of acute shortages have already occurred; at the end of 1973 as a result of bad harvests, drought in Africa, climatic disturbances in Asia and other circumstances, the world food reserves were practically depleted. What will happen in thirty years' time, seeing that development is always accompanied by demands for a better diet? At current yields, the uncultivated reserves of arable land would be unable to meet this increase in demand, at least in Asia (where such reserves are less than 20 per cent), the Middle East and North Africa (where virgin farmland is practically nonexistent). It is hoped that technical advances can intensify agricultural production, but their scope will be limited. Thus, although the 'green revolution' carried out in Asia in the years 1968–70 achieved a considerable increase in the production of certain cereals at the outset, the maximum production levels attained were far below target, and the rate of increase in production was insufficient to match the population growth-rate in the subsequent five years. It is not difficult to see why so many experts are pessimistic.

There remains the possibility of introducing innovations (for example, by developing agriculture), practising greater crop rationalization, and improving storage and distribution channels. Personally, we share the confidence of those who believe that the planet is capable of feeding as much as four times the present human population. On the other hand, we are faced with the disturbing fact of soaring production costs and the resulting inequalities. Thus, the poor peasants of India have great difficulty in achieving a pre-1973 cereal production level on account of the rise in the oil price which puts up the cost of irrigation, and so on. There is no telling where the cycle of impoverishment and inflation which could ensue will stop. We have already reached a stage where excessive consumption on the part of some means that others will be undernourished. What will happen when there are 7,000 million human beings on the earth?

Faced with the need to buy their food, a good number of countries have to step up industrilization with a view to exporting products to pay for food imports. Furthermore, this industrialization will be necessary to provide employment for future generations. The doubling of the population will require the creation of an equal number of jobs in the next three or four decades. In fact, the doubling of the population will call for more than the doubling of industrial productivity. At the same time the quantitative and qualitative needs of the world population will increase, particularly among nations currently in extreme poverty. Formidable problems thus arise. It is clear that the various economic problems are particularly serious, not to mention those relating to the increased energy demand and atmospheric and water pollution. Greater industrialization will be accompanied by a rise in costs proportionate to the increase in demand—all the more so since, at the same time, the depletion and growing scarcity of certain raw materials will enhance their value or involve very expensive synthetic substitutes. In other words, rapid industrialization of the planet, corresponding to the population upsurge, may initially upset the economies of the industrially advanced countries (if the new industrial powers practise dumping, in view of their low labour costs), and lead to a growing and irreversible increase in prices. The Western world is already forced to defend its economy foot by foot against Japanese competition; and, there is also already considerable anger in South-East Asia against Japanese ascendancy. What will happen tomorrow when the Republic of Korea and Indonesia (not to mention Nigeria, Brazil, etc. and, above all, the People's Republic of China) enter the lists in an even stronger position? What do we do then? Can every country be an exporter? Are we not hurtling towards a state of world-wide industrial overproduction which, apart from causing unemployment, may bring about a sharp decline in economic growth in developed countries, or widen the gap between gross national incomes (two eventualities which are not incompatible)? Looking at the question in terms of industrialization, we feel that every nation, no matter how 'wealthy' it is and how stable its population, should feel concerned about the increase in the world population, or at least about its long-term implications.

In reality, industrialization in developing countries presupposes investment by affluent countries. In the present political context, it is hardly likely that these countries will export their capital in any significant volume. Hence, the coming decade will probably see an acceleration in migratory movements, creating a problem which could assume considerable proportions if popualtions double in the poorest countries. By the same token, we are faced with a question which concerns all who dwell on the face of the earth. In 1980, Bangladesh will have 637 inhabitants per square kilometre, the Republic of Korea 431, and Taiwan 473,[10] that is, 1.5 to 2 times more than the most populous European country. If the population of these countries continues to double every twenty-five to thirty years, we must expect a saturation of available space with large-scale emigration. But where will these people go? A formidable migratory movement is already in process: it involves the desertion of rural areas and extremely rapid urbanization on every continent.

According to United Nations statistics, the world urban population, which stood at 405 million in 1925, rose to 1,585 million in the course of fifty years. The figure could in fact be higher, since the meaning of the word 'town' or 'city' varies from country to country. Between 1960 and 1970, the world population growth-rate was 1.88 per cent (2 per cent according to certain estimates), very unevenly distributed among the various countries of the world; the urban population, however, rose by 3.6 per cent with a 2.35 per cent increase in cities with more than 1 million inhabitants. The latter percentage, moreover, should be relativized according to continents: thus, there was a 5.1 per cent population growth in African cities with 1 million or more inhabitants as against an overall population increase of 1.85 per cent. These percentages were, respectively, 3.25 per cent as against 2.2 per cent in Asia; 3.45 per cent as against 2.85 per cent in South America; 1.84 per cent as against 1.55 per cent in North America; 0.85 per cent as against 0.8 per cent in Europe; with a percentage of 2.1 per cent in both cases in Australia.

Experience shows that an excessively high urban density, or an excessive disproportion in the size of the principal cities in relation to the country as a whole, serves to restart the migratory process. In developed countries, the phenomenon is expressed by the emergence of an urgent need to escape from the city (at least temporarily: for holidays or at week-ends). A similar exodus is starting to take place in poor countries; particularly, of course, in urban areas with high unemployment. But the inhabitants of these large cities are unable to return to the outback and they now tend to leave their own countries. Where will they go? And how many such persons will there be in thirty years' time, wandering in search of a place of shelter and a job, from country to country, from continent to continent? An extremely serious question of international law arises in this context: to what extent may countries close their borders, condemning entire populations to extreme poverty and possible extinction (when the individuals concerned have reached the limit of endurance and cannot be held responsible for their fate)?

However cursory and incomplete they may be, these observations suffice, in our view, to show the urgency of the need for a coherent demographic policy. They also indicate the extent to which all nations are jointly involved in this question. While certain nations must restrict their birth rate, others must limit their rates of economic growth and consumption. This solidarity (and the concerted policy which should embody it) should take effect immediately, as the normalization of the world population is a long-term proposition which will take about a century to realize. There are two reasons for this. The first is, quite evidently, the disparity between the various demographic situations. The world population is very unevenly distributed: certain countries have considerable potential resources and a low population density, while others have a high density of population and limited resources. The second reason is that abrupt changes would not only be impossible (a generation would be needed for their effects to be felt), but would also be disastrous if they served only to upset the balance within populations, particularly by creating an excessive disproportion between their economically active and inactive members.

We shall give a single theoretical example.[11] Let us suppose that it was decided to reduce to zero in thirty-five or forty years the population growth rate of a country with 48 million inhabitants, where the birth rate is 6.5 children per woman. It would be necessary to bring the fertility index down from 11 to 1 during the course of this period. However, forty years later, in order to maintain the balance between births and deaths, and between the economically active and inactive members of the population, it would be imperative for the fertility index to rise to 6 (in relation to the population which would by then have reached the figure of 69 million). The school-age population would rise to a maximum of 7 million fifteen or twenty years after the start of the operation, and would then drop to a minimum of 2 million ten years after that period. It would then rise again and finally stabilize at around 4.6 million. If this experiment were to begin in 1980, the economically active population would be around 42 million in 1985, would climb to 75 million in 2030 and would fall back to 42 million in 2070. It is clear that such swings would result in insuperable socio-economic (and cultural) problems.

Whether we like it or not, a sound demographic policy requires gradual restraint of population growth. For the next hundred years, therefore, the world population will continue to increase. But the population increase will be highly disproportionate among different countries and continents. On the basis of the United Nations forecasts, Table 1 may be compiled showing the approximate level and date of demographic stabilization of the various parts of the world.

We should allow for the errors which may creep into this kind of forecasting, and discount any factors which may considerably alter its conclusions (most experts have been surprised by the extent and the swiftness of the decline in the birth rate in Western Europe in the past decades). One thing, however, is

TABLE 1

	Millions of inhabitants	Year
Europe	696	2075
North America	448	2080
U.S.S.R.	445	2085
Far East	1,968	2085
Oceania	57	2090
Latin America	1,609	2100
South-East Asia	4,784	2110
Africa	2,338	2120
TOTAL	12,347	

certain: we shall see an accentuation of the demographic disparities and imbalance between the continents. If we do not wish to let part of mankind suffocate, tear itself apart through war or genocidal atrocities, or lay itself open to various forms of hardship with possibly disastrous consequences, now is the time for us to take the necessary steps towards—and demand from governments—the changes in the world economic system and the level of investment which alone will enable future generations to survive in a state of relative peace and human decency.

On the other hand, economic development is not the miraculous panacea for population problems. There are three reasons for this. In the first place, there are countries where an excessively high birth rate rules out any kind of development. The additional expenditure required each year as a result of the increase in the birth rate claims a large share of the gross national product to the detriment of investment. India and Tunisia are notable examples of this phenomenon. Moreover, although the industrialized countries are today less prolific, their demographic stabilization has come about somewhat late (four generations) in relation to their industrial 'take-off'.[12] Initially, economic development may accelerate population growth; in other words, growth only becomes self-regulating when populations have been 'established' in a state of development for a considerable time. Furthermore, it would appear that this self-regulation is due not so much to economic conditions[13] as to improvements in communications, education and the rise in the cultural level. Lastly, in general terms, it is to be noted that the position of those who maintain that the population explosion should not give any cause for concern since economic

development will keep pace with it rests upon a hypothesis (apart from the fact that it underrates the
questions of population density). Naturally, the pro-
jections of demographers also involve hypothesis. But,
all things considered, their conclusions (at least those
issuing from United Nations sources) tend to paint the
picture less black than the facts warrant).[14] On the
other hand, those who are confident that the combined
effects of development alone can restrain population
growth and meet the enormous needs of a world
population which has at least tripled, overestimate the
development potential of the planet: they disregard the
depletion or exhaustion of certain resources, various
natural limits (for example, the reserves of arable land
gradually encroached upon by urban sprawl), the price
to be paid for improvements in output (particularly in
agriculture) and, above all, the implications (in terms
of costs and resources) of a qualitative rise in living
standards on a world scale. We must repeat that it is
not enough for economic growth to double when a
population doubles, otherwise the present level of
disparity and injustice between nations will simply be
maintained.

For this reason, certain governments, aware of their
responsibilities towards future generations or already
beset by present-day demands, are now linking with
their development activities policy measures aimed more
directly at controlling the birth rate. For this purpose
they use various socio-economic incentives (for example,
by delaying the marriage age, or by limiting family
allowances to a certain number of children), publicize
birth-control methods and try to educate the population
towards an attitude of responsibility.

No doubt we must today take note of the fact that
birth-control techniques are accepted 'both as pre-
conception methods, mostly in developed countries,
and post-conception practices in the developing
nations.'[15] But there is no more than apparent consensus on this issue; profound differences of opinion
remain.

Some relate to the methods used. The deliberate
termination of pregnancy as a method of birth control
is generally rejected (at least, it is not openly accepted)
in traditionally Christian countries, although ethical or
religious objections to the practice are gradually being

overcome.[16] On the other hand, abortions are performed widely in Eastern Europe and other regions of the globe with a different culture. In the same way, sterilization is being (or has been) encouraged by a number of governments, but is not without its share of excesses (pressures, social constraints, the performance of the operations in unhygienic conditions). In other countries it is rejected (by public opinion), advised against or practised with extreme caution, in view of the psychological problems which it may cause and its irreversibility, at least in the case of male sterilization.[17] Even the use of the 'pill' arouses diverse reactions. It has apparently gained unqualified acceptance in certain regions or among certain strata of the population, while it remains suspect in other regions or among other social categories. Admittedly, there are 'pills' and 'pills', and female organisms do not react identically to this kind of contraceptive. Contraceptive pills should only be taken under medical supervision, which makes their use more problematic in undermedicalized regions where they would be most necessary. In this connection, we have no hesitation in condemning the practice of testing certain 'pills' by putting them on the market in developing countries before introducing them in the more affluent countries.

In reality, the 'technical' criticism of these various methods stems widely from cultural or religious objections. Due consideration should be given to the latter. Birth-control campaigns are potentially dangerous without appropriate educational back-up. We should not delude ourselves in this matter. While contraception is justifiable, and even necessary, if couples are to exercise a responsible attitude in regard to their fertility, it may have a disastrous effect on marital fidelity and public morality, with an intensification of the frenzied erotization which has taken hold of Western society. The complete dissociation of sexuality and fertility may lead to an indiscriminate pursuit of pleasure and a 'de-historicization' of desire, both of which induce neurosis. Furthermore, although contraception can contribute to the liberation of women, it often does nothing to make them any the less mere objects of male pleasure. Lastly, independently of the possible disastrous effects, in sociological terms, of a sharp drop in the birth rate, we must guard against changing the family pattern too abruptly by imposing excessively rigorous family-planning schemes. For the individuals or couples concerned, the decision to have or not to have a child is not a clear-cut matter, as it involves a complex assortment of motivations. At national level, the problems are far greater: in fact, just as they are unable to determine scientifically a population optimum, demographers cannot convincingly explain the deep-rooted forces which dictate the steep rises or the decline in the birth rate of a population. The decision to have a child is not merely an expression of the personal motives of a couple: the latter are influenced by a collective unconscious, a 'memory' whose power and causes are virtually unknown. Subtle modifications may occur in a nation which, without necessarily affecting the population growth rate, reflect structural changes in the family pattern. If it is accepted that the child represents the 'will-to-live' of its parents and society, a certain attitude towards happiness, a form of commitment to the future and a rejection of death, a perception of survival, in a word—hope—birth-control measures cannot be imposed indiscriminately. If they were, we would run the risk of imprisoning populations in a state of despair or 'conditioned' hope, of bringing about a civilization without adequate expectations, apart from those connected with the production of material goods. At any rate, the arbitrary imposition of birth-control methods on individuals or categories of individuals against the popular consensus of their nation could only lead to extremely frustrating situations.

In other words, birth-control education schemes are doomed to failure unless they are set in a global context and reflect basic human aspirations. Tradition, culture and religion here play a determining role, creating a field of values regarded as no less essential than those resulting from mere socio-economic advances. Thus, many Africans and Asians question Western development and the consumer society.[18] Actually, any claim as to the desirability of a certain level of development is the outcome of an indeterminate value judgement: the usefulness of such development must be established. Accordingly, it is only natural that the active supporters of birth control should be challenged by those who assume responsibility (by conviction or by function)

for the protection of a certain spiritual heritage or
legitimate cultural aspirations. On the other hand,
governments have the authority to challenge cultural
and religious leaders as to the merits of their ideals,
which should not be promulgated in an abstract,
disembodied way, irrespective of the basic conditions
for the survival of society.

Incorporation and solidarity

One of the points to emerge from any consideration of
the various aspects of this problem is the importance of
national and international solidarity—socio-economic
and socio-cultural solidarity, and also socio-biological
solidarity. Our observations concerning birth control
also apply to the protection of our genetic endowment,
the regulation of human behaviour, etc. We could
survey all the possibilities opened up by contemporary
biology and show that any intervention concerning life
must be assessed, made compatible and controlled in
terms of the common good of mankind, which is to be
regarded as an organic totality with its own set of
conditions for survival.

Such an assertion may appear to be a truism. In
reality, we are far from accepting its implications. On a
cultural plane, at least, particularly in Western countries
we continue to deny society any opportunity to
intervene in the lives of individuals, unless it is to
protect life and to ensure the observance of the basic
human right to live. There is no question of disputing
the basis of the Universal Declaration of Human Rights
(need we recall that society, like any other living
organism, must accept the primary demands of the
cells and organs which constitute it, or else it will simply
undergo necrosis?) However, we should go beyond the
excessively narrow and negative aspects of such a
conception and integrate it in a positive context of
organicity. Every society has to temper the exercise of
individual rights to the common good. Thus, for
example, very young girls are forbidden to marry in
Western society. The purpose of this prohibition is to
safeguard their health (the risk of early pregnancy),
their freedom (as they might experience parental or
marital constraints against which they would be
defenceless), their responsibility (since they might

contractually undertake duties towards their husbands and children beyond their abilities), etc. By the same token, we should accept that society may and should control all biological matters, and not merely procreation. At a time when there is an intensification of economic ties, concentrations, communications and exchanges of all kinds, we do, and shall increasingly, experience the extent to which our distinies are interdependent. It is time that we heightened our awareness of the fundamental reason for this interdependence: we can only survive and attain self-realization through membership of a society which, at present, irreversibly takes on the stature of mankind at large.

Thus we come back to two of the basic propositions stated in the preceding chapter. In regard to life, ethics involves regulating out 'whys' (in this case, our personal needs, wishes and plans) in terms of 'how' (here, the demands arising from our human condition). Hence, we have to accept its authority, in the etymological sense of the word, that is, as a 'guarantee'[19] and not as arbitrary coercion.*

Furthermore, an ethical approach should aim at a 'de-termination' of life, which goes beyond personal interests. We all benefit as individuals or as a nation from the services of other people. It is not only just that we should reciprocate by observing the laws and regulations of society, but also necessary that we should enter into the dynamic process which constitutes the social body, or otherwise suffer rejection. There are those who oppose this incorporation, which, in their view, limits the scope of their lives. Undoubtedly, they have failed to perceive its 'vital' significance: the 'logic' of the living being has unfurled and unfurls in accordance with what appears to be an increasingly elaborate law of organization (inexplicable, perhaps, but real nevertheless). This is the price of the evolution and progress of living beings. We shall have to show that this 'logic' is also the 'logic' of freedom and intelligence.

Accordingly, we may add a further proposition to those which we have already advanced in our attempt to delineate an ethical approach to the problems raised by biology: the 'good' is what strengthens individuals in their solidarity and is directed to incorporation at a higher level.

However, as it stands, this proposition is open to

* It is important that individuals recognize their shared responsibilities with regard to the community as a whole and that responsible authorities increasingly listen to and support the prerogatives of the individual resulting from his awareness of change in scientific advance.— *Conclusion of the paper presented by Dr Verhoestraete.*

criticism insofar as it may be interpreted as denying individuality. We have to show that all organicity involves the preservation of at least the essential characteristics and needs of all the members of the organic body, even if initially such characteristics and needs appear to be divergent.

We should recall that one of the basic functions of society as a whole is to reconcile the 'good' of the individual with the common 'good'. In the field of biology, numerous practices would have to be analysed in this context. As we do not intend to treat this question exhaustively, we shall confine ourselves to a number of brief observations concerning three such practices. Our sole aim is to single out various points which will advance our moral thinking.

Ethical demands

One of the primordial duties of society is, evidently, to ensure the health of all its members to the best of its ability. To this end, it is essential for society to 'guarantee' the harmlessness of the environment (there is no purpose in returning to this question here) and the various products marketed to the members of society. An important aspect of this duty is the control of food and drugs. It is obvious that society is intensifying the compulsory measures applied in this area (and will continue to do so) as individuals produce and absorb an increasing quantity of artificially recombined substances. It is also obvious that these compulsory measures apply (and will continue to apply) to producers and consumers alike. The most recent moves in this connection are aimed at controlling the abuse or misuse of drugs. Nobody would surely dispute the fact that society should reduce the consumption of tobacco and alcohol,[20] not to mention narcotics and hallucinogens. But is public opinion in the more affluent countries prepared to see tighter controls of the sale of pharmaceutical products? The view of biologists, and of various well-informed geneticists and neurologists in particular (we have already referred to this point and will have occasion to return to it),[21] is that the need for supervision in this matter grows increasingly urgent.

* (Such practices)
represent, in fact,
potential infringements
of the right of individuals
to exercise a free choice
because of overriding
considerations regarding
community well-being.
They can be considered
an infringement 'by
necesssity'. Most of the
developed countries have
adopted such measures
in regard to smallpox
vaccination and immuniza-
tion programmes, a
recent example being
vaccination against
poliomyelitis. Depending
on the situation, govern-
ments have used
compulsory measures or
have relied on voluntary
acceptance . . .
. . . Although an
individual is always a
member of a community,
the well-being of the
latter may not necessarily
coincide with the rights
an individual feels are
his own. Compromises
between these two
interests result in a
spectrum of human rights
which reflect the whole
gradient of decision-
making. These range
from decisions resolved
on an individual basis to
those for which the
community or govern-
ment is basically
responsible since they
aim at the protection of
the community as a
whole and by implication
of individuals. In this
respect, individual
members of the
community should be
helped to understand the
underlying compromise
since they themselves are
the ultimate beneficiaries.
—*From the paper
presented by Dr
Verhoestraete.*

'Excessive self-medication has now become a serious problem in developed countries. A large number of persons now exercise their individual right to submit themselves to medication about which they have, more often than not, received insufficient guidance from the medical profession. The phenomenon is the result of present attitudes to pain, discomfort or stress and, often, excessive advertising. Although these are not dependence-producing drugs in the sense of physical dependence, they may in the long run prdouce a form of psychological dependence and their continued excessive use may represent a form of chemical exposure with similar long-term results as those of other known chemical pollutants.'[22]

We should add that in certain cases drug abuse induces new diseases[23] and may give rise to serious accidents if, in the course of subsequent treatment, the medical personnel involved unknowingly prescribe medicines which are contra-indicated in view of the condition of the patient. But the major aspect of this question, about which we are poorly informed as this is a recent phenomenon, concerns the repercussions of such abuse on the progeny of the drug users.

In this context, we may make an initial observation, which is theoretically obvious but disregarded in practice: the well-being of the community is not limited to individuals at present alive; it must also be conceived in terms of future generations.

Concern for the common good also requires the protection of the population against diseases or hazards which are or may be carried by individuals. The most typical example of this form of 'guarantee' is the vaccination programmes. Undoubtedly society has to use persuasion and even compulsory measures in this connection.*

But how far should society go? Some people are allergic to certain vaccines. Also, such practices cannot be applied systematically throughout the population. An example in this respect is the mass screening of whole populations or parts of populations. Such screening may be useful, for instance, in combating tuberculosis, but is it always justifiable? In the light of the various abuses which are known to have occurred, the Final Report of the Varna meeting calls for a certain measure of caution in this connection. Such pro-

F

grammes should only be allowed if there is a proven need for them.*

In this respect, we would put forward a twofold submission, the merits and the ultimate implications of which will only be apparent in the following chapter: from the biological point of view, respect for the common good cannot justify measures resulting in the standardization of individuals or the practice of discrimination. We again appear to be stating the obvious. But we should not lose sight of these basic truths at a time when formidable temptations lie in wait for us, in a society facing an alarming range of questions posed by the cost and deterioration of world health.

We can do no more than touch upon this subject. It is a well-known fact that health costs weigh more and more heavily on the budget of every nation: the consumption of medicines, the number of personnel, the number of hospitals and the amount of hospital equipment are all on the increase.[24] Even taking into account the manifold abuses (excessive medication, useless treatment, insufficiently justified time off from work, etc.), there are so many other inadequacies that we may justifiably conclude that the efforts made in this area by individual countries fall short of what is actually needed. In the coming decades, there can only be substantial increases in health expenditures—not simply in absolute figures, but in relation to the various national incomes. There are a number of different reasons for this: global deterioration, specific disorders due in particular to urban concentration and industrialization, numerous secondary effects which increase the needs of the population, and the development of highly technical new treatments, requiring specialized personnel and costly equipment, which can mitigate certain physical deficiencies, prevent certain forms of suffering or keep alive patients who would have previously died. A whole new area of sophisticated medicine is developing and it inevitably gives rise to enormous social problems. Here, we shall confine ourselves to a few of these problems.

First and foremost, there is the relativization of life and death. Medical practice presents itself as a conditional art: it does not simply comprise an aleatory aspect; it also depends on a number of conditions—temporal (speed of intervention), subjective (if only the

* At the population level, mass screening has been carried out for the detection of heterozygotes of genes causing serious autosomal recessive disease in homozygous situation. Such screening may be useful but should not be compulsory. Moreover, its necessity should be carefully weighed within the different priorities of health-care programmes. The example cited in connection with the mass screening of parts of the American black community in order to discover carriers of sickle cell anaemia showed the far-reaching ethical, legal, social and economic consequences which such procedures may have. The medical authorities, for example, were accused of racial discrimination. Sickle cell anaemia is known to be largely a scourge of people of African origin. Owing to a breach of medical secrecy, a certain number of people screened lost their jobs, social positions, life insurances, etc., being unjustly considered a medical hazard, whereas only their progeny might inherit the disease.— *Extract from the Final Report.*

accuracy of diagnosis) and material (particularly the equipment). Furthermore, it is a well-known fact that every day, in any large hospital, doctors have to take decisions on the advisability of starting or continuing a given treatment. Such decisions are taken in the light of numerous economic, technical, social, professional, scientific and personal variables relating to the patient, his/her family, the medical team, the hospital and even the current socio-political situation.[25]

This relativization has furthered the development of selective medicine, selective not so much in terms of the interests or basic assumptions of the doctors or medical teams as in regard to the relationship of the patients themselves with the medical environment, or in regard to their cultural background and educational level.[26] There is also an increasing gap between the cost of medical-care technology and the low budgetary allocations for preventive medicine, even though the latter may benefit a far greater number of persons. This gap is already considerable in developed countries. It is alarming if we compare the investment in medical care technology in these countries and investment in preventive medicine in certain developing countries.

Whether we like it or not, this increase in expenditure and the various forms of discrimination that are emerging call for the implementation of a comprehensive health policy at national and international level. But the creation of such a policy presupposes a thorough consideration of the strictly ethical implications of these problems.

Such a study is all the more necessary in view of the rapid deterioration of the genetic endowment of mankind. We can mitigate the effects of formerly fatal diseases, although unable to eliminate their causes. Thus, a fair number of individuals can survive and marry. While they do not always transmit their disease to their children, they at least pass on a predisposition or various after-effects which will entail costly medical care. We confine ourselves here to three examples: first, according to certain statistics, the number of diabetes in the world has increased twentyfold over a period of fifty years; secondly, the frequency of genetic diseases is five times higher than prior to medical intervention; lastly, the number of mentally handicapped persons (with varying degrees of mental handicap) equals 2 per cent of the population in France. The fact is that if a generation treats persons with hereditary diseases, it will have to treat their progeny for the same diseases.

It is commonly said that we have abolished natural selection. But we still cannot treat every sick person, carry out every possible treatment or meet the cost of such medical care. Does this mean that we shall have no choice but to reintroduce selection on an artificial basis? Natural selection favoured the reproductively most capable. What will be the criteria of our artificial selection?[27] Dr Verhoestraete concluded his paper with the following observation:

'A major problem for the world at large is perhaps that the social and behavioural sciences have not progressed at the same pace as the natural and biological sciences. As a result, their effects on philosophical and moral thinking, including religious, ethical and religious codes, have been limited. They have, therefore, tended for too long a time to ignore the need for reassessment of the changing values within the structure of modern society. Thus, they have lagged behind in their ability to properly influence the political and social systems of communities and in this way, the direction and application of technological advance.'

Personally, we venture to repeat once more that in our view progress in philosophical and moral thinking presupposes a more thorough grasp of the 'logic' of the living being.

If we enter into this 'logic', we shall more easily perceive that life (individual lives, the life of the community, the life of society) is essentially finite, that it unfurls in time, and that it belongs to this world; and, also, that a right cannot be dissociated from the conditions in which it is exercised, just as freedom must be seen in terms of that from which we endeavour to free ourselves. We incessantly create for ourselves an ideal from which we purport to fashion a reality. Or, to revert to the terms in which we have already outlined the problem, we fix for ourselves a 'why', from which we then proceed to infer the 'how'. But life is not always a definitive given: it proceeds from a dynamic process which thwarts us when we ask the question 'Why?' and in relation to which we have to reassess the aims and rights of society and its members.

If this is the case, we should not be indignant that the reconciliation of the common good and individual good proceeds somewhat tentatively, not without occasional errors: life is a continuous compatibility test, with incessantly re-emerging tensions that are overcome through more intensive control.

Entry into this dynamic process implies recognition of its irreversible nature. Something which has been incorporated cannot be severed without also being destroyed. Every living being is in some way consecrated to a more developed organicity. In this context, we should re-examine the notion of constraint. There are arbitrary forms of constraint, which are intolerable if they lead to a levelling of individuals or discrimination against individuals. In such instances, instead of growing organically, society is in danger of sclerosis or destruction. But there are also constraints which relate to control: and it is obvious that the more the living being organizes, the greater are the repercussions of such control on the future of each of its parts. Moreover, if living beings accept such control, their lives evolve and take on meaning. Biologically speaking, the living being is not its own *raison d'etre*. A light doomed to swift extinction, it exists only as a moment in the dynamic process in which it is involved. As we shall have occasion to repeat time and again, its only purpose discernible at phenomenon level, its sole end and term is to 'de-termine' or go beyond (sublate) itself, giving birth to another living being (we shall return to this point in the chapter entitled 'Reproduction and Sexualization', or incorporating with other living beings, thus participating in an evolutionary process and the constitution of qualitatively higher forms of life.

Sublation and qualitative change are inevitable, otherwise the living being will be rejected by natural selection. We know that down the millenia natural selection has ruthlessly eliminated living beings which were ill-suited to their environment or more vulnerable to predators. But such selection operated only *a posteriori*. It was not selection which led to the qualitative change, hence evolution, of living beings, but vice versa. If as we aspire to an understanding of the 'logic' of the living being with a view to gaining a certain mastery of life, we have no choice but to pursue, and help others to pursue, this obscure travail of qualitative change. And, once again, this depends primarily on the possibility of organic incorporation in the social body. For this reason, it is evident, for example, that we should not bring into the world an excessive number of children or beings that will clearly suffer from extreme debility, lest they cannot be incorporated and are inevitably smothered or rejected. Individual good entails constant reference to the common good. Conversely, however, there cannot be respect for the common good without respect for the good of individuals.

Notes

1. It would appear that the main reasons for the disappearance of these monsters were the excessively slow reactions of their nervous system, the absence of reflexes, coupled with the malfunctioning of certain relay nerve-centres, rather than climatic conditions or a food shortage.
2. Sickle cell anaemia, for example, is common among blacks.
3. Various participants referred to this question: Professor Gajewski, Dr Der Kaloustian and Dr Narayan. Furthermore, a considerable part of the paper presented by Professor Dubinin was devoted to it.
4. Paper presented by Dr Der Kaloustian.
5. Final Report.
6. More especially as it will be possible to fix the new species thus obtained by means of cloning (the reproduction of identical copies of an individual organism—particularly through parthenogenesis—in order to avoid the disparity which may arise from sexual reproduction). It remains to be seen whether this new rationalization of agriculture applied in crop farming, does not involve the risk of disturbing the ecosystems to the extent that its use is generalized and tends to reduce the number of species cultivated so as to concentrate on high-yield crops. Once again it will be necessary to compensate, but at what cost?
7. A number of international conferences have already been held on this subject, notably in London in 1970 and New York in 1971, the latter being attended by eighty-five scientists from six countries.
8. So much has been written on this subject that it is impossible to give even a summary bibliography here, more especially as the arguments advanced are sometimes contradictory. The following pages are based on the conclusions of United Nations studies and the gist of articles which we ourselves have contributed to *Études* (July 1974 and August 1974).
9. At Varna, this topic was discussed principally in the paper presented by Professor Dubinin and the report by Dr Mroueh.
10. Obviously we are omitting any mention here of the cases of Hong Kong, Macao and Singapore, which are 'city-States'.
11. These observations are based on the analyses of J. Bourgeois-Pichat and Si-Ahmed Taleb, who took as the starting-point for their study the demographic situation of Mexico in 1960 (an article entitled 'A Zero Growth Rate for Developing Countries

by the Year 2000: A Dream or Reality?' which appeared in the journal *Population*, No. 5, 1970).

12. In the nineteenth century, countries with a European culture, where the fastest industrial growth took place, saw their annual birth rates almost triple, rising from 7.6 million in 1800 to around 20 million by the end of the century.

13. Although the living conditions of a highly urbanized and industrialized population (particularly where female employment is widespread) place very considerable constraints upon the birth rate.

14. They predict a population of 12,700 million in 100 years' time, whereas a straightforward projection from current growth rates gives a figure of around 30,000 million.

15. Final Report.

16. This is an established fact and does not imply any value judgement on our part. We shall return to this question in the chapter entitled 'Reproduction and Sexualization'.

17. In fact, in highly medicalized countries (particularly those where there are sperm banks), there has been a sharp increase in the number of voluntary sterilizations, involving couples who do not want any more children, and couples who carry a serious hereditary disease. We shall return to this question in the chapter entitled 'Reproduction and Sexualization'.

18. This is a further argument against the view that economic development automatically leads to voluntary birth control. Does not such a view unjustifiably project Western forms of behaviour on to the world at large? There is no evidence that, given the same standard of living, an African or an Asian would behave identically to a Westerner. The case of Japan is far from convincing, in view of the peculiar geographic and cultural situation of the country and the considerable influence which the West has exerted on it in the industrial and military spheres.

19. From the Latin *auctor*, meaning 'guarantor'.

20. There are those who call for an absolute prohibition on tobacco, or at least on its sale to minors. They seem to forget that a simple prohibition could itself lead to very grave social abuses (illicit trade or black-market operations), that such prohibition has shown itself to be ineffectual and that it sometimes results in the sale of even more dangerous substitutes. This was seen in the United States during the prohibition years.

21. cf. the chapters entitled 'Problems and Implications' and 'Reproduction and Sexualization'.

22. Paper presented by Dr Verhoestraete.

23. cf. the chapter entitled 'Problems and Implications', observations of Professor Dubinin.

24. It is difficult to assess health-care costs. Many expenditures cannot be included in national statistics, especially those paid by individuals in the context of private medicine. *A fortiori*, reliable international comparisons are out of the question, even in regard to highly medicalized countries. In France, total health-care expenditure recordable at national level rose from 24,000 million francs in 1968 to 69,800 million in 1974. 'Calculated at the 1968 value of the franc, per capita medical consumption went up 63.8 per cent in eight years, that is, 6.4 per cent a year. In terms of the present-day value of the franc, the increase was 16.9 per cent'—P. Longone, *Population et Société*, No. 91, May 1976.

25. We shall return to this theme in the chapter entitled 'Senescence and Death'.

26. Various studies have shown that the more affluent and better educated classes are less prone to certain diseases. This may be

explained by the fact that they have a healthier diet, avoid certain errors in the event of physiological accident, or have a certain insight into the nature and gravity of their illness at the time of its onset.

27. At the present time, when doctors are confronted with urgent problems (for example, if they do not have the equipment or personnel needed to carry out resuscitation), they favour patients with the best chance of survival, the youngest patients or those with the greatest family or social responsibilities.

Individuality and adaptability

The good of the individual

To respect the primordial demands of individuals is a vital obligation for mankind; it would court disaster if it were to try to even out all differences. This statement is obviously based on the fact that, like all organisms, society is made up of separate cells whose development or functioning cannot be disturbed without jeapordizing the survival of the whole system. It also reflects the fact that organicity is always the result of complex interactions, no single one of which can be suddenly or arbitrarily suppressed or modified without setting up chain reactions the effects of which cannot be foreseen. But the statement also rests on even more compelling considerations, pertaining to the very core of what we know about living beings.

Preserving differences

No living systems is known which is not both individual and 'open', even if a number of 'repressors' prevent anarchic differentiations and mutations. What can be observed of the organism as a whole also holds good for each of its component parts. It is impossible to analyse here all the 'mechanisms' or the possibilities which enable a living being not only to distinguish itself from others and undergo transformation, but also to adapt to circumstances and develop. Let us give only a few examples.

One of the basic reasons for this individuality and this 'opening' probably lies in the characteristic molecular structure of living beings. There exists in each of them a 'force which introduces *asymmetry* in chemical activities and cannot be imitated in the laboratory'.[1] We now know that this 'force' originates in enzyme activity, and one of the major objectives of biological research is to discover how its effects may be controlled. This is a matter of the utmost importance, as the asymmetry so introduced is the key to the cell's 'nutrition'[2] (and therefore to its conservation, its growth, its release of a quantum of energy needed in order to overcome entropy, etc.). It is impossible here to go into all the implications of this activity, but these basic facts cannot be evaded: life is an unceasing process of decomposition and recomposition; to be sure, the recomposition is rigorously programmed and regulated, but the fact remains that life presupposes a dynamic appropriation peculiar to each cell—and this is a constant that is confirmed at all stages and at all levels; organicity integrates and confirms diversity, even if this means increasing the number of regulatory mechanisms.

Every living system is therefore a chemically unique being, even though it is subject to identical internal necessities and an identical 'logic'. We know that this diversity originates in the mixing of genotypes. The millions of genes arranged along the chromosome chain (each of them controlling one or more enzymatic or other functions) are redistributed, half by half, at conception, as the chromosomes of the male and the female are recombined (and because they, too, are the result of a combination, the probability of two individuals being completely identical is practically non-existent after tens of thousands of generations, except in the case of twins). In addition to these innumerable combinations, the effects of mutations and 'genetic drift' must be borne in mind: the first result from changes in individual genes, the second from an increase or decrease in their numbers. Thus every living being probably receives, as basic endowment, a very great number of genes which are typical of the species or group to which it belongs and a large number of 'markers' that have characterized the biological peculiarities of its parents. However, it also inherits genes which have had no obvious influence on the previous generation but which, when they are recombined, may have a decisive effect on its own constitution;

lastly it may inherit mutant genes or a surplus or deficiency of particular genes, whether for good or ill, either equipping it with new qualities or faculties of adaptation, or causing serious, and even mortal, malfunctioning or disequilibria.

At first sight, this individuality of living beings might be taken for a weakness. It is, indeed, one of the prime factors of individual morbidity; however, it is also in reality one of the fundamental mechanisms safeguarding the species. For convincing proof of this it is sufficient to consider (with some simplification of the processes involved) the speed at which microbes thwart the calculations of the producers of antibiotics. A weapon believed capable of wiping them out is discovered and shortly afterwards they reappear, apparently unsusceptible to its effects, and new weapons have to be forged. This resistance is due more to selection than to a sort of immunization or transformation induced by the antibiotic, although this is also possible. In the teeming population of a species of microbes there is always at least one intractable individual that will survive and soon recreate a new race;[3] further, it may even be able to 'contaminate' some of its fellows and transfer the same resistance to them.[4]

An 'abnormal' microbe thus sometimes becomes the saviour of its species. Similar phenomena occur in human beings. In certain cases, individual peculiarities, even pathological ones, may confer a sort of immunization against more dangerous diseases. In this connection we have the example of sickle cell anaemia (drepanocytosis) which confers on the black people afflicted with it an exceptional resistance to malaria.

These succinct reminders were necessary to situate more clearly some of the problems which are now facing biologists—and were at the heart of the discussions at Varna.

It is very tempting to pursue research to neutralize biological individuality, notably by intervening directly on the genetic endowment of individuals. Studies to this end are, obviously, to be encouraged when they aim at overcoming certain diseases or defects, or at counteracting rejection phenomena in the case of organ transplants. But, even if it were to become possible on a large scale, it would be wholly wrong to combat apparent genetic anomalies thoughtlessly or to seek to

develop systematically certain 'qualities' of living
beings.

Generally speaking, as Professor Gajevski pointed
out: 'it is only possible to speak in a relative way of the
inferior qualities of genes or markers, as it only makes
sense under strictly defined conditions of life. We have
no way of knowing whether such genes as we now
consider inferior will not be of great value in the
future, either combined with other genes or under
different conditions which the species has not yet
experienced . . . It now seems probable that, whatever
the conditions, genes which are lethal or sublethal[5]
in the homozygous state may have positive effects in the
heterozygous state.[6] It is, therefore, hard to say that
one gene is less useful than others under present or
future conditions. Any project for eradicating lethal or
sublethal genes from the human population is impractic-
able. Each one of us possesses and transmits some ten or
fifteen lethal genes . . . which are present in all our
chromosomes, and new ones appear when genetic
mutation occurs. Some of these may be important for
the future evolution of mankind.'

The importance not only of structural genes but
also of regulator genes in the evolutionary process is
obvious if these considerations are examined in greater
depth. It is known, for example, that the chromosomal
chain of the chimpanzee corresponds 90 per cent to
that of man; so, apparently, only 2 per cent of genes
are responsible for the specific difference, and the
greater part of these appear to be regulator genes.
That is to say, the entire regulatory system is not only
the means of achieving homeostasis but by its very
variations it is also one of the principal agents of
evolution.

Thus, it is not surprising that biologists with a
sense of responsibility are reticent to extend genetic
manipulations to man and that, *a fortiori*, they reject
the prospects opened up by cloning.

This word is applied to a series of techniques which
make it possible to avoid the hazards of heredity
(notably by parthenogenesis) and to fix acquired
characteristics so that the living creature to be born
will be absolutely identical to the parent—or may even
be endowed with original properties which would not
normally be transmitted. Research in this field is

based on a desire to create by artificial means plant and animal species with very high yields or adapted to particular living conditions. These species are of tremendous importance in view of the prospect of the world population's being doubled or tripled.

However, what may be the long-term results of the use of this technique even in the case of plants? Is there not a risk of starting an escalade of artificiality which could certainly provide mankind with additional resources, especially food, but could also lead to certain crops being developed at the expense of others? In the case, for example, of a climatic catastrophe, will these privileged species survive? Would not a wider diversity of species offer greater hope of detecting more resistant ones? An easy rejoinder is that the ability to clone would make it possible to create at will other species adapted to the new conditions. This is so. But it should be added that a lot of time will be needed before the species so created proliferate in sufficient numbers. Let us imagine, for example, that a disease suddenly affects a certain variety of wheat and rapidly kills it off. It would be possible to recreate the varnished species artificially and immunize it, but the number of plants would be limited. It would be necessary to wait for the 'harvest' to gather the seed these plants would bear—and how many years would thus pass before enough were available to sow a hectare of land, and then to supply seed to the whole world? Now, such disasters are not unprecedented: one has but to call to mind the ravages which phylloxera caused in vineyards, or the way certain varieties of oysters which formerly stocked French oyster beds have practically disappeared. If it had not been possible in both these cases to turn to other varieties the disaster would have been irreparable for several centuries. Within the 'logic' of cloning it is not impossible to imagine the whole earth covered by two or three species of cereals and about twenty species of trees, the other plants being confined to 'natural' parks or specialized gardens. This would result in the variety of genotypes being extremely restricted. And it also no doubt implies that a no less dangerous disturbance of all ecosystems would be set in motion, raising manifold problems in connection with crop rotation, water cycles, decrease in fauna, and so on.

Nevertheless, bearing in mind that the earth will have to feed 12,000 million human beings, it would be just as dangerous to forego the advantages which cloning provides. 'This procedure has already been successfully accomplished in plants as well as in amphibians. If mammalian cloning could be achieved, it would be utilized to great advantage in livestock. Experiments in this direction are bound to continue.'[7] It is to be hoped that the results of this research will only be applied with caution.

'Once cloning in animals is achieved, that of humans will become probable.[8] What would be the legitimate uses of human cloning? 'Certainly the general speculations of having either "supermen" or producing people to do unwanted jobs or for human experimentation, with replaceable parts available for convenient transplants, are insufficient to motivate this kind of scientific research. They are not only insufficient, they are naive. A human being is more than his genetic potential. The interaction of his genetic variables with the environment, to a considerable extent, shapes the "person". The technical steps necessary to do human cloning are likely to be inspired not by the quest for a super-race but by the need to solve compelling problems.'[9]

These 'compelling problems' were not clearly defined at Varna. They could arise either from a sudden transformation of the environment (following some disturbance in the solar system), or, alas, in a less fanciful hypothesis, from a large-scale contamination of mankind by bacteria or by atmoic irradiation, disasters such as would leave only a small number of survivors who could reproduce 'healthily'.

But leaving aside these extreme cases, even if one can foresee its being technically possible in the near future (by *in vitro* chromosomal duplication using two cells obtained from the same individual, with subsequent reimplantation in the uterus), any other use of cloning, as well as the forms of eugenics already mentioned, should be rejected. It has been said that, for the common good, it will be necessary to 'preform' individuals as has happened in the case of certain animal species such as ants, or on the lines of what already occurs in all complex living beings whose every organ has its own attributes and functions. Supposing that such specialization were possible (though, once again, it is not technically speaking possible in view of the fact that, as we have seen, we do not comprehend the working of the genetic system as a whole, with its interacting functions and controls— and even if we did, this would not alter the fact that the individual is also moulded—made the individual he is, by virtue of all his relationships), such specialization would be detrimental to society, for two reasons. First because 'heterogeneity with a species is compulsory for the proper selection of the fittest. As cloning implies uniformity, it would, biologically speaking, be an unsuitable process and would result in evolutionary weakness of the species'.[10] Secondly, and above all, because heterogeneity or, if one likes, individuality, even at the biological level, opens the road to qualitatively higher forms of existence. This point deserves to be somewhat amplified.

Individuality and emergence

There are no doubt living species which have undergone practically no transformations for millions of years. However, the fact remains that throughout history— and forming part of history itself—life has undergone an evolution leading to the appearance of more and more complex, organized and mobile living beings, possessing a nervous system which has allowed them to accede to forms of intelligence which are themselves evolutionary. Has evolution stopped, as some people say it has? We have no grounds for supposing so, especially as man has disturbed the balance of the plant and animal kingdoms, hunting and wiping out some species while remodelling others at will. It is possible, for example, that the companionship between men and certain animals such as the dog (apart from artificial breeding practices) may, by changing their way of life, result in deep-seated transformations which only time will bring to light. It is also possible that, under identical appearances, improvements to the central nervous system are occurring in certain animals and even in mankind (particularly owing to the intensification of social relationships). Nevertheless, evolution towards better-adapted forms of life, not to speak of higher ones in the sense of their being more complex and more relational, cannot be taken as a purely inconsequent phenomenon. Even if we think of this evolution as being the result of chance, the fact remains that the possibility of attaining an enhanced state of being is inherent in living creatures, virtually written into their genetic programme, and that it must be safeguarded as a 'reserve' for the future; not only to guard against a possible natural or artificial transformation of the environment, but because obstructing evolution would, as it were, stop the living being's time sequence, whereas physical time and the time of thermodynamics would continue to unfold; in fact, as we shall see further on, to immobilize the history of life would be to condemn it to death.

More immediately, if it is necessary to keep intact the possibility of 'emerging into' an enhanced or fuller state of being, the reason is clearly in order to ensure the good both of the individual and of society. Who then would dare to determine or predetermine the existence of the generations which follow us? What 'preformation' would be the most necessary, bearing in mind changing individual and collective needs)? As Professor Gajewski said, 'We have no objective criteria for appraising which human characteristics are positive and should be strengthened and propagated. It is all a question of personal opinion . . . Is intelligence really more valuable than a refined emotional life, or artistic gifts? These characteristics are all equally precious. Each human being is a veritable treasure-house, and this has a deep biological meaning.'

These remarks also apply to society, which is basically evolutive and needs to be constantly improved. Who is able to discern all the mechanisms and inter-locking relationships involved in the structure of society? Who can draw its history to a close or decree what its destiny is to be? We most certainly have to perfect its organicity and, with this in view, seek to improve its condition. But it would be a mistake to equate improved condition once and for all with fuller being, thus making a value judgement which will inevitably be incomplete. Whether he likes it or not, man must resign himself to being contingent, alive, caught up in a state of flux which eludes him. Once again, for him improvement—the *better*—does not necessarily imply greater fullness—the *more*—and the *why*, expressed by value judgements or utility judgements, must be a function of a *how* formulated in terms of opening up and emergence.

This last observation shows how essential it is to relativize the concept of abnormality and, above all, to curb the common temptation to weed out the abnormal (defined with respect to what criterion?). Under the pretext of combating the abnormal do we not actually run the risk of impairing what is individual? At a genetic level, we have seen that even lethal and sublethal genes may possibly constitute a 'reserve' for the future; or that a genetic illness may be an antidote to other diseases. Abnormality should be thought of in terms of how the individual is affected by his state, whether he can or cannot survive, and how he fits into society. This, in turn must be understood both negatively, insofar as the abnormal individual endangers the equilibrium of society, and positively, in as much as he may be a contribution to society, an addition to the 'programme', inciting it to react organically in a new way, to discern and to adopt more suitable forms of existence.

This was what Professor Vartanian had in mind when he said at Varna: 'One day, one of my French colleagues said, somewhat jokingly, that if we were to adopt a policy preventing mental patients from becoming pregnant the world would die out in dreariness. As in all aphorisms there is a hint of exaggeration here, but there is also an element of truth. The fact is that it has been convincingly proved that heterozygosis is bound up with a whole series of advantages, and particularly with a high level of intelligence, and if we look at this problem from the standpoint of evolution, we may say that these illnesses are the price humanity pays for its multiform nature, its polymorphism'.

This leads us directly to one of the most difficult and preoccupying problems facing contemporary biologists, that of the modification of human behaviour. Although we are theoretically convinced of its being necessary to preserve the uniqueness of each individual, both for his own good and for the good of society, in fact we constantly act in such a way as to infringe on his reason and his liberty (to mention only these two faculties). Before outlining the position in regard to some current practices, and in order to perceive more

G

clearly how to regulate them, it might be wise to recall briefly some elementary facts concerning the relationships between reason and liberty on the one hand and biology on the other. For, at the very time when our contemporaries as a whole are becoming more clearly aware of these relationships, we act without paying sufficient attention to them; we determine, or seek to determine, physiological traits without worrying about the possible ensuing effects on our mental faculties (and we have pointed out how vital it is that people who are preparing to undertake genetic interventions should take these psychological effects into account); conversely, we try to stimulate our psyche artificially without worrying about the repercussions these stimuli may have on our organism.

Although many of our contemporaries know that their brain is an extremely complex system, very few of them have an inkling how very singular it is. Certainly all individuals have approximately the same brain structure and in all probability they all possess about the same number of brain cells (although it is hard to state anything definite on this last point, as the number of neurones has been estimated at several tens of thousands of millions). Incidentaly, as a human being has about 2,000 million genes it would seem obvious from the outset that the brain is more complex than the genetic system itself. Once these common characteristics are accepted, however, analysis reveals the individuality of each brain at all levels. First at the level of the neurone which, like all the cells of our body, constantly produces proteins and enzymes. It has now been proved that the latter are not exactly identical; not only do they differ from one race to another but even from one individual to another. Furthermore, each neurone is connected to others by millions of tiny filaments, the dendrites, through which it receives and transmits information like a computer. However, apart from having a greater number of circuits, the neurone has the peculiarity that, unlike even our most highly perfected machines, it recreates these circuits continuously. The dendrites are in perpetual movement, new connections are made, old ones terminated.[11] Let us imagine that it were possible to isolate 2 million neurones in ten individuals, that these ten 'systems' were chemically and functionally identical, that these neurones could be numbered and that one might observe, for instance, neurone No. 20 in each of them. We would see that neurone No. 20 of system 5 was not connected to the other cells of its own system in the same way as the corresponding neurone of system 6. The latter might perhaps appear to be connected directly to neurone No. 21, whereas this connection would not exist in system 5. If we were to repeat this observation some time later we would see that all the connections had changed and that one now existed between neurones 20 and 21 of system 5, whereas the one we had seen in system 6 had 'come undone'. This is of course a gross over-simplification: as there are tens of thousands of millions of neurones, each connected by millions of dendrites, it is literally impossible to conceive of the number of constantly changing combinations. In other words, even if our brain were chemically and structurally similar to our neighbour's it would not function in exactly the same way.

There is an even more disconcerting fact. This connecting and disconnecting activity undoubtedly complies with certain laws and affinities. This was proved by Roger Sperry in 1940; he enucleated the eye of an anaesthesized toad and grafted it the other way round, whereupon each of the nerve cells of the retina rapidly remade its previous connections with the brain cell to which it corresponded. In other words, these cells 'found' one another again, a stupefying example of recognition.

Thus a structure was reformed, and probably the first 'concern' of our nervous cells is to constitute and preserve these structures in a way which is practically identical for all individuals. However, secondary affinities and recognitions also exist, which appear to be specific to each individual.

In addition to this system of actions and reactions there is a complex set of controls of all sorts, which ensure the 'good' functioning (activities, retroactivities, stimulations, repressions, etc.) of each of our neurones and of the entire brain. What is more, the two hemispheres of the brain 'control' each other. This is a new factor which partly explains the intellectual, emotional and behavioural singularity of each individual. In this connection, Professor Deglin's paper, transmitted to the participants at the Varna meeting, was very instructive.[12] Some lengthy extracts from this paper are reproduced below:

'One of the specific characteristics of the human brain is what is known as the functional specialization of the cerebral hemispheres. It has been discovered, literally in the last few years, that the left hemisphere controls logical and abstract thinking, whereas the right controls concrete and imaginal thinking. The personality and modes of perception of an individual depend on which of his two cerebral hemispheres is more developed (whether as a result of inherited characteristics or education).

'. . . If neither the dimensions nor the weight of the brain are the distinguishing characteristic of the brain of "homo sapiens", what makes it unique? Today there is only one characteristic of the human brain which we can regard as unique, namely its functional asymmetry.

'The brain of all animals and of man himself is symmetrical; its right and left halves are constructed identically both in regard to the composition and quantity of their individual parts and in their general architecture. In animals, the right and left halves perform the same work. In man, however, the right and left cerebral hemispheres have different functions and govern different kinds of activity. . . . Today, functional asymmetry is emerging as perhaps the main scientific problem in connexion with the human brain.

'. . . There is a definite principle behind the functional asymmetry of the brain viz., the left hemisphere governs logical and abstract thinking, whereas the right hemisphere governs concrete and imaginal thinking.

'. . . Accordingly, each hemisphere, each apparatus, has its own set of instruments; its own speech, its own memory, its own emotional tone.

'. . . The child is born with what may be regarded as two right hemispheres and has as yet no "verbal" hemisphere . . . Occupational specialization of the hemispheres in human beings is completed after birth and a demarcation line between the apparatuses for imaginal and for abstract thinking becomes established with increasing age. Moreover, it appears that a person's individuality and psychological make-up depend on which of the thinking apparatuses becomes dominant.

'. . . The two hemispheres are not therefore independent of each other. There are complex and paradoxical interconnections between them. On the one hand they cooperate in the work of the brain, each complementing the abilities of the other; on the other hand they compete as though each were preventing the other from doing its own job. . . . In the nervous system, stimulation is always accompanied by inhibition. . . . Without this inhibitory process the activity of the nervous system would become chaotic, undirected and self-destructive.

'. . . But the mutually inhibitory effect of the hemispheres has one further specific purpose. In order to react satisfactorily to the changing circumstances and varied situations with which the individual is faced in his everyday life, it is essential sometimes to combine the aptitudes of the right and left hemispheres and sometimes to use the capacities of one or other of them to the fullest possible extent.

'. . . So that it should be possible to use the apparatuses for imaginal and abstract thinking individually, it is essential that they be separated from each other and be situated in different parts of the brain so that an intensification of certain abilities do not entail an intensification of the others . . .'

These observations on the brain raise a question as to the nature of intelligence. This is no place to formulate a theory on this subject, neither is it necessary to show how essential it is for an individual's survival (conservation, reproduction and adaptation) once it has reached a certain level of organicity; nor how it conditions all our interpersonal relations, and therefore our integration into society, without which we could no longer subsist. For the moment we shall restrict ourselves to five propositions which, although they are well known, nevertheless should be constantly borne in mind at a time when we are tempted to multiply interventions on our nervous system and standardize our behaviour.

First, whatever the specificity of the human mind may be,[13] the brain is obviously not simply the vehicle or instrument of thought, it also plays a part in shaping thought. We must beware of anthropomorphisms, but there is no denying that animals possess a form of intelligence and that man displays a certain arrogance when he claims to be a creature completely 'apart'. He has feelings, perceptions and basic behaviour patterns which other living beings certainly had before he appeared in the evolutionary chain, although these 'impressions' and forms of behaviour are modulated differently from one type of organism to another.[14] What is true of individuals is also true of their relations with their fellows.[15] In other words it seems that, in living beings with a brain, there develop progressively patterns of intelligence and, hence, behaviour which are originally identical. It would be desirable for contemporary biologists, ethologists and sociologists to undertake a concerted analysis to understand these patterns, as it is evident that if they are disturbed in any way, this results in serious psychic trauma which, quite apart from the suffering it causes, may handicap the individual, as well as in behavioural deviations which make it difficult, or even impossible, for him to fit into society. To put it yet another way, there seems to be a sort of 'common trunk or core' (an 'order' or 'logic') of intelligence which, of course, branches out differently from one species to another (notably as a function of their brain morphology), but to which the individual must remain attached. From this point of view it is legitimate to modify the psyche or behaviour if one or other should go off the rails, legitimate also to intervene on the nervous system if these deviations result from its malfunctioning.

Second, however, it is equally obvious that as (because) life itself is polymorphic, so intelligence is polymorphic too. This is easily seen in regard to species, but is also true of the individual; everyone has his own specific biological individuality, particularly on a genetic and cerebral level. Consequently, not only is it fruitless to attempt to standardize the manifestations of intelligence or to render behaviour patterns uniform, but to attempt to do so would result in very serious personality disorders.

Third, we must, on the contrary, preserve this individuality of intelligences, for the sake of the common good itself, insofar as, to mention only the field touched by biology, it is a factor conducive to adaptation, incorporation into society and the enrichment of social relationships; and also insofar as this intelligence may give rise to qualitatively superior forms of existence and transmit their benefits to society.

Fourth, one of the most disturbing aspects of life is exactly this surprising diversity of species and individuals, some of the genetic and cerebral aspects of which we have already mentioned. (We could have called attention to many other manifestations if we had, for instance, invoked the hormonal system or the histocompatibility complex.[16] It is a fact that the finer the biological analysis, the more it finds the unique in the common, the different in the seemingly identical. It is thanks to this that not only adaptation, but also evolution and emergence become possible. It is as though a living being always retains within itself the potentiality of fuller being.

Finally, perhaps some people wish to see this evolution cease, thinking that, biologically and socially, we have reached the *nec plus ultra* of existence? If this were so, we would confine ourselves to a state of purposeless finitude, and soon end up in sclerosis. We must ask ourselves what makes this evolution biologically possible, the *how* of emergence, in short the purpose of life. But it is, we believe, already manifest that one of the major imperatives confronting mankind is that it should beware of standardization and preserve the individual in the organic. And this imperative has an ethical force.

The modification of behaviour

Although these considerations are very general, they emphasize the importance which we must attach to the research now being carried out on changing or manipulating behaviour. This is a tremendous subject which was alluded to, rather than dealt with, at the Varna symposium,[17] and which, perhaps more than any other, should be approached on a transdisciplinary level. It is closely connected with ethology and sociology on the one hand and with the entire field of psychology and neurology on the other. And such a subject is dealt with differently depending on whether it is considered from the point of view of society or from that of the individual (and a distinction must also be drawn between the relatively healthy individual who uses or receives a variety of stimuli on the one hand and the mental patient on the other). We can, therefore, only touch on it here, simply recalling certain fundamental points in the field of biology without broaching a number of related problems pertaining to ethology and sociology.

Far be it from us to think it reprehensible *per se* to modify human behaviour. First, since, as we have constantly said, human beings are fundamentally relational, and only realize their full potentialities in society, they should clearly be given every assistance in acquiring or developing this relational character and fitting into society. Secondly, as Jacques Bril's paper so strongly emhasizes,[18] not only is every species charac-terized (or even constituted) by the anatomical, physio-logical or morphological properties of its members, but also by a certain number of specific features of behaviour directed towards ensuring its continued existence and identity. The same thing applies to mankind.

'Man has a wide range of behaviour patterns, as though nature had transferred the coercive character of instinctive, i.e. unquestioned, forms of behaviour from, innate, genetically transmitted structures to acquired sociologically taught ones.

'. . . These structures have the function of main-taining the existence of the group as a solid entity, of ensuring, as it were, its ontological solidarity, while allowing individual aspirations partial satisfaction. Each society, each culture, thus possesses a large number of these internal regulating instruments designed to control individual drives while drawing its own energy from them.

'. . . This fundamental, unitary aspiration of the species to maintain its identity underlies countless modes of cultural production whose diversity, far from testifying to plurality, on the contrary reflects the ingenuity with which the species is capable of taking advantage of the most varied environmental stimuli or limitations.'

Society has always provided itself with numerous means, starting with education and teaching, of influencing the behaviour of individuals. Their use has been directed 'at will towards the enhancement, restoration or destruction of mental structures, whether individual or collective'. Nowadays, however, these means have become considerably more efficient, both for better and for worse. New forms of action on behaviour have appeared, the use of which needs to be judiciously controlled. They may be classified under two headings: those which tend to modify the individual's relations with his circle, and those which directly affect physiology.

The first may be divided into three groups (often intermingled), according to whether they are directed towards 'inundating' the patient with stimuli which will finally sweep him along in the desired direction (an obsession or a desire is then created artificially); triggering an aversion reaction; or 'desensitizing' the individual to certain aspirations, demands or obsessions. These techniques have been used in many ways over the centuries 'in numerous educational, religious, military, political, industrial and other contexts, generally accepted—even if locally contested—by a balanced society'. Nowadays they are increasing considerably, thanks especially to the achievements of psychoanalysis (though not without sometimes distorting these or using them for questionable ends). 'There is no "communication" technique, no style of management, no sales method, etc., which does not in one way or another involve the deliberately angled use of a body of knowledge originally sought for therapeutic ends.'[19] The danger inherent in this type of procedure is obvious when any particular authority having adequate means of communication at its disposal is able to 'amplify certain sociological processes vis-à-vis an individual or a group placed artificially in a confined environment. Some specialists go so far as to envisage the elimination of all socially "undesirable" behaviour and the advent of socially "compatible" responses by creating a behavioural and psychic environment around the individual or group'.

As Jacques Bril also says, it is true that 'the sociological structures engendered by the dynamics of evolution comprise self-defence mechanisms', particularly of an ethical, cultural or religious nature. Nevertheless, the specific characteristic of these methods for modifying the relation of an individual to his social environment is that they act on these self-defence mechanisms, and in such a way that the thresholds between attraction, incitation and constraint are often, and purposely, difficult to discern or define; particularly is this so where these methods seek to bring about contingent effects. We can, obviously, lay down two 'principles by which we may define the legitimacy of efforts to manipulate psychic behaviour:

'Any manipulation which does not aim at safeguarding or at restoring the identity of the individual should be prohibited as being an attack on the biological integrity of mankind.

'Any other motivation which does not include one or other of these objectives, explicity or implicitly, is tantamount to a destruction of mankind's immunological defenses.'

Nevertheless, these principles, which are more easily applicable in the case of an intervention on the physiological level—we shall return to this point further on—have only very limited scope where techniques aimed at changing human relations are concerned, bearing in mind the ambiguity of such methods in regard both to their goal and to the means they adopt. For this reason, Jacques Bril continues:

'Besides extremely valuable work carried about by various commissions, such as those set up by the C.I.O.M.S.,[20] which are engaged in differentiating between the means and circumstances for which manipulation would be legitimate and those for which it would not, studies must also be conducted on the real motives lying behind the intervention and the criteria used for detecting these motives.

'Nevertheless, mankind's means of self-defence against the risks it brings upon itself are only circumstantial palliatives: these formidable ills can really be prevented only by educating people to an awareness that they are all members of one species.'

But have we not repeated often enough, awareness of belonging to one species, or better, awareness of the true nature and ultimate purpose of society, necessarily entails recognition of the fatal danger that threatens society if individuals are made to conform to the same pattern and their individuality and liberty reduced.

This danger exists today. It is the result not only of the manipulations we have just mentioned, but also of direct interventions on the physiology of the nervous system, even though carried out for therapeutic reasons. They are of three kinds.

First, psychosurgery. This word covers 'all surgical operations which may be carried out on the brain. It is capable of changing behaviour by inducing a lesion of limited extent in a precise region of the brain. Of these operations lobotomy is the best known to the general public; it dissociates certain regions of the cortical zone from the diencephalic affective centres with which they are normally connected; or roughly speaking, it dissociates specifically human behaviour from the instinctive behaviour that precedes it. A large number of more or less specific interventions have now been developed such as resections, electrocoagulations and various cauterizations used to calm hyperactivity, alleviate anguish, curb sexual appetite, and so on.'[21]

It must be admitted, however, that the surgeons who undertake this type of operation do so, for the most part, without really knowing what the consequences will be either at the level of treatment for the disorders they hope at least to alleviate, if not cure, or from that of the lasting psychological side effects which may result. Hence the misgivings of the majority of neurophysiologists with respect to psychosurgery.*

Certain diagnostic or therapeutic techniques involving electrical or radio-active stimulation of the brain centres, are allied to psychosurgery but are less dangerous because they are not irreversible. Electroshock therapy is best known of these interventions. New methods are being progressively developed, some directed towards refining this type of operation by stimulating only a very localized area of the brain, and others tending to replace electric current by radio waves. 'Stimoreceptors' have been invented that are capable of picking up both the manifestations of cerebral activity and, conversely, of 'informing' the brain by appropriate waves. We probably cannot expect this type of stimulation to serve as a substitute for psychic activity; it is unable to create thought or language. However, it will provide valuable information on the nature of mental troubles or on the effects of medicaments prescribed to treat them. It may palliate a particular deficiency arising from the malfunctioning of one or other of the cortical zones, thus making it possible to cure certain forms of deafness, for example. It will also make it possible to regulate or activate functions of the brain, setting up or removing certain inhibitions, as in the

* New methods exist which have a limited therapeutic effect while causing a minimum of harm to patients; we are now in a qualitatively new era of neurology. I must point out, however, that a group of experts who analysed this information came to the conclusion, first, that the results were not so very beneficial, and second, that side effects and complications were not so uncommon as they were said to be. All things considered, I believe that although we have gained experience in treating disorders such as Parkinson's disease and many forms of epilepsy, the problem still remains as delicate as it was . . . Certain neurosurgeons . . . under the pretext of treating patients are really only carrying out research when they undertake such operations.—*Extract from Professor Vartanian's communication.*

case, for instance, of one cerebral hemisphere dominat-
ing the other too strongly, as Professor Deglin explains
in his communication mentioned above. Similarly,
some specialists envisage moderating excessive emotional reactions, and particularly aggressiveness, to the
extent that they reveal physiological deficiencies. All
things considered, these techniques will certainly
provide new means for modifying human behaviour
and the relations of the individual with those around
him.

As they require neither apparatus nor installations,
chemical treatments are much more widely used. 'They
may relieve much suffering and induce more freedom
of the psyche, but they may obviously also give rise to
considerable alienation by causing a possible loss of
will power, instinctual deviation and a manipulation of
moral sense.'[21] As Professor Vartanian pointed out:

'Like all sciences, psycho-pharmacology contains
many contradictions. On the one hand it has radically
changed the entire face of psychiatry and of nervous
disorders, making it much easier to keep patients in
their own surroundings rather than in a hospital, and
it has been of invaluable help in the treatment of
mental disease. On the other hand, biologists and
psychiatrists are extremely concerned about the amount
of psychotropic drugs people obtain and use. Society
should be as alarmed at this state of affairs as specialists
are . . . What is most surprising is that 70% of all
prescriptions for psycho-tropic drugs are not made out
by psychiatrists or neuropathologists, but by general
practitioners or interns—in other words by people who
are not qualified in this field.

Naturally, such a situation may have completely
unforeseen consequences for mankind. We do not see
in our clinics the nervous and mental troubles and the
very serious complications arising from the widespread
use of these products, as the firms who sell them and the
medical and governmental organizations responsible
for testing them keep careful watch to prevent it. Also,
we do not know how these drugs act, we do not understand the working mechanism of any of them. We do
not even know how aspirin works—despite the fact that
all sorts of people take large amounts of it—to say
nothing of complex psychotropic substances. We all
know perfectly well that, throughout its evolution,

the brain has never been faced with anything like this problem. Our brains have been beset by such a vast quantity of diverse drugs only during the last twenty years or so. . . . Evolution has not provided the brain with systems of enzymes to react specifically to these substances, and as it is impossible to foresee what form they might take, it is not possible for us to assess the repercussions of all these substances. . . . Even worse, we know nothing of the consequences these drugs may have . . . on future generations, the descendants of the people who have taken such quantities of drugs.'

We have thus given ourselves powerful means of intervening as much on the physiology as on the behaviour of individuals. For better or for worse. Needless to say, to safeguard the individuality and liberty of each human being, we must be more on our guard than ever to protect people from undue inter-ference on the part of the group or society, and also to preserve society from the vagaries of individuals. It is doubtful, however, whether control from the outside can ever be completely effective. A whole new concept of education is needed, one that will lead to a rigorous ethical code.

Ethical requirements

Promotion of this new concept of education and code of ethics is a matter for society as a whole. In particular those in positions of responsibility must be made to feel more conscious of the fact that the common good, even when it postulates a more rigorous organicity, implies respect for the individuality of the person, inasmuch as on this depends the possibility of progress and emergence.

It is true that the task of combating disease and suffering, as well as that of safeguarding the biological potential of individuals, is becoming more and more of a burden upon society. Over and over again we have come up against the inescapable opposition existing between the common good and the good of the indi-vidual. Everything leads us to believe that their opposi-tion will intensify and that no regulations laid down *a priori* will enable the ensuing strains to be overcome. What is needed is a compromise reflecting a national consensus and therefore veering to one side or the other according to each nation's level of development. And in order to arrive at a valid compromise, and a valid policy giving effect to it, it is essential that not only the people responsible for treating and taking care of patients but also those who seek their help should be more aware of the respect due to the individuality of each and every one, bearing in mind his or her psychology and socio-cultural environment.

Owing to the lack of medical facilities,[22] and often to lack of time and various economic considerations,[23] there is a great temptation to standardize forms of treatment. We do not wish to analyse or criticize medical and hospital practice here. It must be admitted that not only does every doctor have his own way of receiving patients, differing moreover according to the country, the culture, the way in which the profession is practised, the specialization, and so on, but also that considerable progress has been made in the last few years to improve human relations in hospitals, although each hospital and even each service has its own specific character.[24] Nevertheless, from the moment an individual steps through the doors of a consulting room, and above all those of a hospital, he is only too often deprived, as it were, not only of his liberty (and necessarily so) but seems no longer to exist as an individual. He is identified with his illness, and becomes 'obstruction of the bowels', 'coronary' or 'cancer'. This propensity of the medical profession[25] to dissociate patient and illness, and concentrate on the latter rather than on the former, no doubt arsies partly from the necessity for a doctor to safeguard his liberty as a clinician by maintaining a certain reserve and partly from a very personal defence reflex on his part. (Is it possible for a human being ever to dominate completely the anguish brought on by illness and the prospect of death?) Probably it is indispensable, and inevitable, that particular cases should be generalized. How could medical science have made progress if the 'laws' of each particular illness had not been defined? And how could one combat a disease if in every case a therapy had to be reinvented? This tendency to generalize may lead, however, to practices which ultimately can appear ethically suspect. All too often, defenceless patients, not really conscious of their state, are subjected to forms of treatment[26] which in the last analysis have no other objective than to gain better knowledge of the complaint and to be of service to medical research. Such practices are perhaps necessary for the advancement of medical science and to save other lives, so should not be condemned in the abstract. But it would be reassuring to know that the 'chief' who subjects his patients to such practices does not really do so to obtain personal knowledge, mastery of his subject and professional renown. We must understand that illness, particularly for a specialist, can become his possession, his reason for fighting—and even for living. This is only natural, provided that the patient is not reduced to being considered merely as the case or instance of this illness, which becomes the only thing worthy of consideration.[27]

In many cases, in fact, the decision to undertake a particular treatment depends on the techniques and apparatus or chemotherapy available. That many patients would not be cured but for these tools should be brought home firmly to those who inveigh against the cruelty entailed in certain forms of treatment. There is, however, another side to this technical progress. We have already expressed regret at seeing the patient treated as a mere instance of a particular illness; we must also deplore the fact that the illness itself, too often, becomes the occasion for technical virtuosity, a pretext for using what, even for the doctor, are the fascinating machines now available in modern hospitals.

Here again we are probably caricaturing a little; this 'technologitis' is more of a tendency than a reality. It is nevertheless true that with the technical equipment now available (particularly for intensive care), providing as it does completely new means and making it possible to reduce the number of accidents or to compensate for failures or loss of equilibrium, the treatment of disease appears to be progressively losing its specific character: with such a wide range of possibilities at its disposal, there is a tendency for the medical profession to fit a particular complaint into this or that form of treatment, or an assortment of technical procedures, rather than to adjust the technique to the complaint.[28] Then what about the patient, one may ask? The answer appears to be subjected to 'deindividualization' twice over.[29]

In the same context, and we have no hesitation in

repeating ourselves in this, as it was one of the leitmotifs of the Varna symposium, we must once again denounce the propensity of many doctors to prescribe massive doses of drugs indiscriminately (sometimes without being sufficiently well informed as to their properties, or without possessing the necessary competence), disregarding their side effects on patients. This holds good for everything from antibiotics to tranquillizers. We have already outlined the dangers inherent in this type of practice as much for his descendants as for the patient himself. We have also seen that the manner in which some doctors conceive the abnormal is open to criticism, particularly where psychic or behavioural disorders are concerned. There is no doubt that any diagnosis is subjective to a certain extent, and that the criteria often applied to the abnormal are not confined to biology alone but also include cultural or ideological prejudices. It also cannot be denied that, in many cases, the disorders we are talking about are in fact due to disturbances in the individual's relations with his own body or with society, and that these disturbances are themselves induced by a preference for certain social and family, cultural and ideological models. Before starting treatment doctors, particularly psychiatrists and neurobiologists, but also psychoanalysts, should question the deep motives which prompt them to prescribe a certain treatment, especially when deciding whether a patient is to be confined to an institution.[30] It would also be well for the patient's family circle and friends to be brought into the process.[31]

To sum up, both in order to arrive at this consensus we referred to as necessary so as to reconcile the contradictory demands of the common good and the good of the individual, and in order to lessen the abuse of medication (an abuse often provoked by patient demand, especially in regard to excessive use of drugs), one of the fundamental tasks facing the medical profession today would seem to be to contribute to this process of education which as we have seen is so indispensable for the future, helping people enter into responsible acceptance of their own biological and behavioural individuality.

A good number of our contemporaries are losing the sense and taste of their own individuality. They are warped by political centralism, pressure towards conformism at the hands of educational and training institutions, economic strategems which employ depersonalizing forms of publicity, not to mention ideological or religious 'models'. This can be seen in our relations and, consequently, in our behaviour; we accept standardization and dislike being different from others, or not having what they have, much more than we wish the contrary—that others should be the same as we are. This may also be observed on a biological level, in that we tend to 'standardize' our bodies to an ever-increasing extent. Apart from the harmful side effects of our interventions, we only too often disregard the fact that if our faculties of reasoning and our liberties are peculiar to us, the obvious reason is not only that our relations with the people we are in contact with are distinctive, but also that our biological conditioning is unique. A great deal of physical suffering as well as anguish would be reduced and many unsuitable forms of treatment would not be undertaken if we would accept being what we are biologically, inasmuch as this also contributes to moulding our personality. There is certainly no question of preaching a kind of stoicism here, but before subjecting our bodies to all sorts of hidden oppressions, it would perhaps be wise to inquire into the nature of the 'ills' which beset us. Insofar as they are an obstacle to our exercising our intelligence or our liberty, and prevent us from participating in the life of society, everything possible should be done to lessen or overcome them, but the fact remains that we shall never completely conquer suffering and infirmities.

We must oppose mercilessly the ruthless side of this 'logic' of life where it leads to distressing disparities in the fate of individuals, but we must also understand the obscure dynamics of the process, insofar as this disparity in fact proves to be the necessary condition for emergency to a higher level. Here we are again faced with the major questions we have already discussed, which lie at the heart of the relation between biology and ethics; at the level of the individual or that of society, can man hope for, and should he in fact aim at, a state of biological 'perfection'? And what would constitute this 'perfection'? Although we shall have to face up to these questions in our last chapter, we can already give a partial reply. Over and beyond the fight against

H

suffering and the quest for greater well-being, for the common good and the good of the individual alike it is essential that, even biologically speaking, the human being should be respected and should respect himself as bearing features which may contribute to the progress of mankind. This presupposes that he deploy, and be helped to deploy, all the potentialities inherent in his individuality. In other words that he style his life—to the end of giving it. In this way it takes on purpose; at least this is what we shall attempt to bring out in the following chapters.

Notes

1. François Jacob, commenting on a remark of Pasteur's in *La Logique du Vivant*, op. cit., p. 252. (Author's italics.)
2. Owing to this asymmetry a new element can be 'attracted' into the molecular chain.
3. Under optimal conditions, it takes fifteen and a half hours to obtain 1,000 million bacteria from one particular 'specimen'.
4. Some living cells have particles that are independent of their nucleus (which houses the constituents imprinted with the hereditary code). These particles 'carry' fragments of coding material which can thus be transduced from one living cell to another.
5. A lethal gene is one capable of causing death.
6. By homozygotes is meant two individuals having an identical gene, or a gene which has undergone an identical change; in the opposite case they are said to be heterozygotes.
7. Final Report.
8. ibid.
9. ibid.
10. ibid.
11. These 'breaks' are all the more necessary as neurones, unlike other cells, do not replace themselves. We all 'burn out' more than 25,000 neurones each day.
12. This communication was published in the *Unesco Courier*, January 1976.
13. We shall return to this subject in the last chapter.
14. It is obvious, for example, that animals' senses are infinitely more developed than those of human beings.
15. The work of K. Lorentz or K. von Frisch should be recalled here; they reveal surprising analogies between the structures of life in society among animals and men.
16. The mutual compatibility of tissues in the same organism or between different organisms.
17. It was to be developed by Jacques Bril, who was unable to be present at Varna, but whose paper was distributed to the participants.
18. Basing himself on the work of K. Lorentz, N. Tinbergen and K. von Frisch.
19. 'Without mentioning the indiscriminating use of group techniques, where the manifestations accompanying the relationship (bodily contacts, cries) are often heightened but at the same time the unskilled leaders are unable to control the emotions, frustrations

and conflicts which may be induced by this artificial heightening of manifestations isolated from their social context'—Jacques Bril.

20. Council for International Organizations of Medical Sciences.
21. Jacques Bril's communication.
22. This is prevalent both in developing countries and in most developed countries, bearing in mind not only the new demands but also the trauma induced by some forms of industrial and urban development.
23. The more uncommon and singular a complaint, the more financial sacrifices are accepted for treating it. This is the case in certain operations which require the most advanced medical techniques (as in the well-known case of heart transplants), whereas such sophisticated means are not used to treat more widespread serious disorders. This is to say nothing of the slender means allotted to preventive medicine.
24. To a great extent this depends on the personality of the 'chief'. It should also be noted that inasmuch as patients and their families suffer psychologically, they tend to relate everything to themselves and fail to recognize that the people caring for them are also men and women. Members of the medical staff are treated according to their function: only very few patients or their families think of a nurse, for instance, as a person with her own psychological, family or professional problems.
25. Except, perhaps for nurses, nurses' assistants and serving women.
26. For example successive blood samples, forms of medication, etc.
27. We shall return to this question when we speak of the doctors' desperate efforts to ward off death.
28. And not only 'complaints'. For example there is a growing tendency to practise reanimation on all babies at birth, whether they are in danger or not.
29. In the chapter entitled 'Senescence and Death', we shall return to the question of his 'right' to participate in the decisions concerning the treatment he is to undergo.
30. This brings up the whole question of the present-day controversy in regard to 'anti-psychiatry'. We do not consider ourselves competent to analyse this subject; moreover, it was not discussed at Varna.
31. It also goes without saying that doctors who indulge in practices designed systematically to standardize the behaviour of so-called 'patients'—in fact to change their mental make-up—for political reasons or on behalf of the police should be strongly condemned.

Reproduction
and sexualization

The concept of good
in one's relation to others

The giving of life. Here we come to what is probably the fundamental characteristic of all things living. With matter, there is a continual reshuffling of its components; with life, there is reproduction. And the whole existence of the living being appears to be organized for this purpose. Thus François Jacob wrote; 'What can possibly be the aim of bacteria? What do they seek to produce that justifies their existence, determines their organization and subtends their work? There seems to be one reply, and one only, to this question. What an individual bacterium seeks endlessly to produce is two bacteria . . . The whole structure of the bacterial cell, its entire more of operation, all its chemistry, have been refined to one single end, that of producing two organisms identical to itself, in the best possible way, as quickly as possible, and in the most varied circumstances.'[1]

Reproduction not only governs the morphology of the living being—its organicity, functions and internal control system—but it has constituted, and continues to constitute, one of the mainsprings of the dynamics of evolution. It was therefore inevitable that man should try to obtain mastery of it, acting under pressure from both individual aspirations and powerful motivations of a social, cultural or economic nature. Today, biological research is largely directed towards such mastery. We have already given some brief examples, reviewing the state of the art as regards genetic intervention, cloning, artificial insemination, the choice of a child's sex, or birth control as part of population growth control. But the progress of science now provides many other possibilities, which call for reflection of an ethical nature.

As the subject broached here is a vast and highly complex one, we can do no more than make a start on such reflection, not with a view to an exhaustive study (this was not possible at Varna) but, as in our earlier chapters, in order to identify a number of signposts which together will enable us to find our way through the maze. In so doing, we must not lose sight of the need to strike a balance between individual good and the good of society.

The sexual 'revolution'

In order to situate the intrinsic ethical requirements in this connection correctly in relation to the 'logic' of life, it is in our view first necessary to recall some elementary data.[2]

With non-sexual beings, reproduction operates on the principle of identity; a single genetic programme is transmitted; any mutations are random, and are only adopted if they are consonant with the continuance of life and resistance to the environment. In addition, the individual divides up for ever: at the level of the bacterium, there is no partner, reproduction is nothing else than the continuity and quantification of the same, which, barring accidents, does not disappear; a bacterium has, as it were, neither past nor future.

It is true that, even chemically, cases occur of molecular compenetration; that a bacterium can be 'impregnated' by agents of a viral type, and that there even exist bacteria known by analogy as 'female', since they appear to be more 'disposed' to this kind of penetration; and that there is a dense network of intercellular information within an organism. Sexualization has its archetypes. Nevertheless, in its most highly developed forms, sexual reproduction (a better way of putting it would be 'procreation') differs radically from bacterial duplication. Its product is not two from one, but one from two. Since there is a combination and reshuffling of two genetic programmes, diversity in identity becomes an 'essential' principle. The individual then survives only through the instrumentality of another individual. Thus three basic elements emerge:

The quality of life becomes more important than quantity:[3] it is the strongest and finest individual, the 'best' from the reproductive point of view, that usually carries the day; and the old gives way to the new.

Sexualization leads to a continual process of sublation. One's survival in one's progeny depends on an opening up towards one's sexual partner, but also, as a corollary, towards the other within oneself, by bringing into play individual potentialities. Thus new types of organicity and control are called for: it becomes necessary to be able to 'live together' even though different, since any kind of self-sufficiency has been radically done away with. The living being is seen as the locus in which life acquires qualities and becomes richer, the matrix from which other living beings develop. It is the channel—transient, self-sublating and, as we shall see in the next chapter, mortgaged to death—through which life is carried on.

Thus with sexualization, life experiences a genuine revolution, since once the living being can no longer normally reproduce itself on its own, it no longer 'determines' life. The determination of life comprises a relationship between two living beings. What reality can we attach to a binary relation of this kind? Yet the characteristic of being able to relate to another individual (not only, as before, to the environment as between the parts) produces as it were a polarization of organicity. The input to the genetic programme includes not only pleasure and desire, but also a host of aptitudes for perceiving or emitting stimulatory signals (olfactory, auditory, visual), without which sexuality would fail in its purpose.

It must therefore be recognized that sexualization introduces a disconcerning element into what we call life. We noted earlier[4] that life does not exist in the abstract: only living beings exist. Yet we now discover that there runs through living beings a 'logic' of ex-sistence, entailing both sublation and de-termination of themselves.[5]

From these brief comments there emerges a preliminary conclusion to which we shall only refer in passing, since it goes beyond the framework of biology as such. If the foregoing observations are correct, it should be noted that, with sexualization, the relational character and the acquisition of qualities play a part of paramount importance. It is not in order to reproduce itself that a flower is all the colours of the rainbow, or has fragrance; but the fact that it is so means that it is better at reproducing itself (and that it will therefore be selected for these qualities). In other words, it is not the state of being sexualized which has created attraction between living beings: on the contrary, it is this attraction (even if only in terms of simple chemical reactions) which has gradually brought about sexualization. To put it yet another way, sexualization is relative not only to reproduction (reproduction existed before sexualization) but also to the qualities possessed by the progenitors. If this is the case, it is not possible to claim that the meeting of bodies is directed solely towards procreation; it should also be seen as a principle of compenetration, of reciprocal exchange, of enhanced value. In this sense it is possible to dissociate the meeting of bodies from procreation. This throws new light on the problem of contraception. However, it appears that from this point of view the exercise of human sexuality should be patterned on desire, whereas in most cases all we have is the desire for pleasure: a pleasure which is short lived and joyless. This is a contradiction in terms, both physiological and psychological, since it is the strength of the desire which creates the intensity of pleasure. And the strength of the desire can only come from apprehending what it is that distinguishes one's partner. In this sense, sexuality is an approach to the out-of-the-ordinary. It is in fact so in its very principle, if we take as its archetype the example quoted above, of the penetration of a bacterium by a virus: this produces, through an exchange of information, a breaking away from their particular order, a shifting of their specific needs, which may result in a new qualification of the one by the other, giving rise to an evolutionary process. It is, however, obvious that such penetration is destructive (the virus either kills the bacterium, or is rejected by it), if there is no reciprocity in the exchange. A desire for desire (the love of love): without this shared act, sexuality is devoid of meaning, hence of morality.[6]

Whatever the case may be as regards the ultimate

significance of sexualization, it emerges from the foregoing that the right to procreate is a fundamental, grounding right, inherent in the very survival of the individual, but that this right cannot be exercised without due respect for the vital requirements of the child to be borne and those of the society, since individual procreation is the reproduction of society, and since the child must take its place as an organic member of that society.

The making of life

The recognition of the individual's right to procreate implies doing everything possible to enable him to exercise that right, hence that appropriate measures should be taken when necessary to overcome sterility, including the use of artificial insemination. We have seen[7] that the Varna symposium recognized this, even when it is necessary to have recourse to AID (artificial insemination by a donor), provided that various precautions are taken in relation to interpersonal problems, the couple and society. Similarly, once it can be done without risk, it will be possible to have recourse to *in vitro* fertilization.*

This technique can be used to overcome sterility in women when this is due to tubal defects. It consists in taking an ovule from a woman's follicles, fertilizing it artificially, possibly leaving it to develop for a short time in a favourable environment, and then transferring it back to the mother's uterus. It is a fact that teams of scientists have succeeded in 'cultivating' embryos on placentas to a fairly advanced stage: for example with a rat embryo, up to mid-gestation.[8] At the present time a human ovum treated in this way has been successfully developed up to the stage where it becomes necessary to implant it in the uterus.[9] What holds scientists back is the possibility that such fertilization might lead to chromosomal anomalies. When this point has been cleared up, and if there proves to be nothing in it, an attempt will be made on the last stage, namely the implanting of the ovum in the uterus of the woman from whom it has been taken.

We are of course speaking of research, and research linked in varying degrees to pathological deficiencies

* *In vitro* fertilization of oocytes followed by early development of the embryo and then transfer to the mother's uterus has been the subject of much recent scientific discussion. Such techniques have been successfully used in certain mammals as the mouse, hamster, rabbit, pig, sheep and cow. Their use in humans may just be a matter of time. It will then be possible for women with blocked fallopian tubes to become pregnant using their husband's sperm. However, such manipulations could disturb the normal process of gametogenesis, fertilization and embryonic cleavage. They could lead to chromosomal aberrations in the developing embryo and thus result in birth defects. With further animal experimentation, however, it may be possible to avoid these complications. I do not think that this type of procedure, when practised to treat certain types of sterility, should meet ethical objections, especially if the methods are perfected enough. The arguments declaring normal procreation as human and *in vitro* fertilization as inhuman cannot be accepted any more, since man has already adopted many different types of artificial means in different areas of his life.— *Extract from the paper by Dr Der Kaloustian.*

However, confidentiality should be maintained by doctors in the interest of patients so treated.—*Extract from the Final Report.*

which scientists are seeking to palliate. But the very fact of putting such research on record—and there are other examples to which we might have referred[10]—gives us the opportunity of stressing two considerations which we regard as of even greater importance.

First, it must be observed that fertility is becoming increasingly dissociated from the sexual act. Since what today constitutes research will tomorrow be common knowledge, and even more common practice—even if such types of intervention concern only pathological cases—it follows that people will then look on the 'normal' sexual act as only one means among others of fertilization and procreation, or rather that fertilization and procreation will increasingly fall into the category of activities which can be artificially brought about.

Second, there is no getting away from the vital question, what exactly is an ovum which has been fertilized *in vitro*? Is it 'already a human being'?[11] Which of us would unreservedly subscribe to such a view—at least so long as the embryo has not yet been transferred back to the mother's body? We have here 'living matter', a being which is potential or *in posse*, but which is not in a state of entelechy, i.e. possessing its end within itself. This distinction calls for an analysis in terms of its philosophical and ethical implications. In the first place, what right have we to draw a distinction between 'potentiality' and 'entelechy'? On the answer to this question depends the solution to many problems relating to abortion and experiments with life. The reason is that if we admit this distinction, it will then be argued not only that there is no 'human being' so long as the foetus does not possess its end within itself.[12] (though as we shall see, this in itself would not provide a justification for abortion) but above all that it is the role of the procreators to 'make the embryo live' as a 'human' being—in other words, that the true nature of the embryo is dependent in part on the intentions of conscious living beings, and is relative to their consciousness. Scientists working on *in vitro* embryos may have no intention of making them live as human beings. Many people consider that the 'making of life' is one of the highest and noblest responsibilities recognized by mankind; others, however, question the justification for such a claim, which is contradicted by currently accepted ethical standards, at least in the culture of the West, imbued with a particular interpretation of Christianity. Our view, backed up by findings about sexualization, is that the very 'logic' of life leads to integrating its relational character into its definition, irrespective of the chemical processes by which life is governed, and even irrespective of the physiology of organisms.[13] There is no human being without a relationship to another human being.[14]

This categorial statement, which will be amplified throughout the rest of this study, will in our view allay the serious misgivings aroused in some quarters by the cultivation of human tissues. It is well known that a considerable amount of current research (medical, biochemical, biocellular) makes use of embryo human tissue cultivated *in vitro*. It is for example this kind of 'stock' which, in view of its singularity, has made it possible to produce vaccines against German measles, and to carry out reliable tests on them. We are all indebted to such research. At the same time, it provokes an outcry in many quarters. Let us not be hypocritical about it: it is the lives of a large number of children which are at stake.

It is possible, if not probable, that in the (near?) future this type of stock will be obtained *in vitro* and experimentation and intervention on 'living human matter' will take place on an increasingly large scale. An eminent member of the scientific community stated in the following terms one of the questions with which research workers are now faced, a question which in turn gives rise to many others:

'As a result of the teratological disasters produced by certain medicines (and it is possible that more subtle modifications have escaped observation), the idea has been gaining ground that all forms of medication should be avoided for pregnant women, chiefly during the period in which the principal organs are constituted. Though one can see good reason for this attitude, it is misleading and sometimes inapplicable. Thought must therefore be given to the possible use of human embryos cultivated *in vitro* for the purpose of testing medicines. This is a somewhat alarming prospect, but in considering the problem should we not approach it the other way round? Would it not have been better to carry out tests on a small number of human embryo *in vitro*, rather than produce 10,000 phocomeles due to thalido-mide?'[15]

Making use of life in the service of life: this is the great adventure in which a growing number of research workers are participating, motivated by the demands on all sides from men and women who, individually and collectively, refuse to be at the mercy of the hazards, disturbances and deviations by which life is beset. It is foreseeable that the techniques thus applied may give rise to abuse. How are they to be assessed, controlled or guarded against? In our view, by making progress in joint reflection on the requirements inherent in the physiological needs and the relational character of human life.

The right to procreate

It is from this dual standpoint that we have to appraise the exercise of the right to procreate. For even if, as we have seen, the exercise of this right is obviously inalien-able, a constituent part of the life of individuals, it can be regulated by a society which is concerned for the free development of both sexual partners, the existence of the child to be born, and its own weal as a social body.

The first requirement is to preserve the genetic heritage. There is no need to revert to this here, except to denounce once again drug abuse and the excessive consumption of tobacco and alcohol. The risks involved about which the public at large is not sufficiently informed, except in certain dramatic circumstances, are now the subject of rigorous study and controlled experimentation (as we have just seen, in the case of tissue culture). The fact remains that it is impossible to guard completely against them. Here two questions arise: can one, and should one, prohibit procreation by those suffering from certain diseases who are obliged to undergo treatment likely to disturb the development of the embryo—for example, women suffering from schizophrenia who are being treated with pheno-thiazines?[16] And what should be done about people who are bearers of serious hereditary defects?

We have already come up against similar problems in connection with population growth (and the increase in the cost of public health borne by the bulk of the population), and we concluded that society was entitled to regulate the exercise of the right to procreate. At this juncture we must go into this conclusion in greater detail. It is clear from our foregoing comments that procreation is a right shared with the other sexual partner and with society as a whole; and that this right should be exercised in a manner consistent with the underlying significance or logic of sexualization, which is to impart a new quality to life and enhance the relationship with other individuals. The deliberate calling into life of a physically defective being, who would be destined to great suffering and would be incapable of developing a relational character and being integrated with society, is out of the question.

Realization of the fact that the right to procreate is shared with society implies a change in outlooks. This change is now beginning to take place, though not without provoking keen resistance, which is moreover justified, as we have already noted.[17] It must, however, be acknowledged that in this area, society has its own specific and inalienable requirements. All organisms control the development of their parts. We should not complain at action to prevent the growth of any one of our organs getting out of control, or to neutralize the reproduction of a necrotic cell. Individuals must advance towards a greater awareness of their organicity, and recognize that organicity connotes as many duties as rights. It is not diminishing the human personality or infringing its freedom if the individual is urged to 'integrate with the body of society' in all the dimensions of his being. Life becomes richer by becoming regulated. Here those responsible for different types of culture (or morals or religions) should examine thoroughly the specific function of that which it is their duty to safeguard. For while it is true that culture and traditions inherit a vital 'memory', this memory is in no way hard-and-fast: the specific 'memory' of each living being has been built up progressively as these beings have become organized into 'integrons'. Today it is necessary that a 'memory' be constituted, and come into play, for the body corporate of society.

Conversely, the requirements of society must be relativized. First, in terms of synchrony, which is unavoidable:[18] for example, sterilization imposed as a form of contraception, if not freely accepted, may generate a conflict of 'memories' which will have a serious traumatic effect on the individual, whereas a decade or two from now this risk may fade out with a new kind of acculturation. Next, in terms of reversibility: it should in fact be noted that an organism controls certain functions, and neutralizes them as necessary, though without on that account abolishing them. To perform an irreversible act, particularly as regard contraception, cannot be anything but an exceptional measure, adopted solely in the case where individuals acting of their own free choice find that it is impossible for them to do otherwise; or where those who are bearers of serious diseases which can be transmitted hereditarily consent, for the good of society, to waive their right to procreate. Even then one must define what is meant by a serious hereditary defect: for while an organism tends to neutralize rather than eliminate, it is because (to adopt an anthropomorphic view) it appears to be noncommittal with regard to itself, as if careful not to alienate the wealth of potentialities latent even in its own disorders (we have in fact seen that some genetic 'diseases' can be used as antidotes for other illnesses). If the dysgenic effects of a particular disease can be tolerated by the sufferer and by society, society has no right to eliminate them by prohibiting the individual in question ever to procreate. On the other hand, society can take 'selective' measures to dissuade two individuals who are bearers of the same disease from marrying or from procreating, if their union would inevitably, or with a very high degree of probability, produce a child which would be substantially and irreversibily handicapped.

It will thus be seen that reproduction calls for thorough consultation and joint planning between society and individuals, a sharing of responsibilities designed to respect the specific good of both alike. We need to bring about a greater awareness of the problem. From this point of view, however, a new ethical question arises: How far should parents be warned in cases where the transmission of a hereditary disease is probable, though not certain?[19] Would this not mean condemning them to an anxious period of waiting (which might last for years, in the case of an illness which declares itself relatively late on), continually on the lookout for advance symptoms of the illness in the child? Yet the converse is no less alarming: to keep parents ignorant of what may happen to their children, let them organize their lives as if nothing were the matter, may have serious consequences, especially if the parents produce several children who all fall ill one after the other. In such circumstances, the author's personal view is that the truth cannot be kept back. But as a counterpart, it seems self-evident that society must help the parents to cope with the responsibilities which will devolve upon them. This implies taking a whole set of measures to support them, ranging from special employment conditions (in particular for the woman) to housing and the provision of welfare assistance, or action to exempt them from certain constraints or obligations.

Therapeutic abortion

If one recognizes the parents' right to the truth about the state of health of an unborn child, does this not inevitably mean envisaging the possibility that they may be led to ask for a termination of pregnancy in the case of an affected foetus? There is no doubt that the possibility of such requests will increase with the steadily growing practice (which is a desirable one) of consulting a doctor in advance of and during pregnancy,* and as techniques for making an intra-uterine diagnosis become more sophisticated.†

'If a serious genetic disease is diagnosed, a so-called "therapeutic" abortion can be resorted to after consent of the couple concerned, with a view to alleviating the psychological and social burden to family and society.'[20] A somewhat vague statement, in which the accent is laid on the consent of the couple, since 'the experts were unanimous on the weight of responsibility borne by the parents in such cases.'[21] However, 'in this context they also referred to the hesitation or refusal encountered in certain countries and among certain spheres of society'.[22]

It will be seen that the subject was touched on at Varna rather than discussed, the real reason being the shortage of time and the absence of a sufficiently detailed document to be used as a basis for thorough discussion.[23] But the absence of any such document was significant. Could it have been otherwise in a meeting of

* Genetic advice can be used as part of marriage counselling or after marriage in order to avoid the conception and birth of children with serious genetic diseases. The couples involved should then be left to take their own decision free of any psychological or legal pressures.—*Extract from the Final Report.*

† Among the different technical procedures used for intra-uterine diagnosis are amniocentesis, sonography and amnioscopy. The last two procedures are still, to a great extent, experimental. Amniocentesis, however, has passed during the last decade from the investigative stage to the stage of inclusion in the medical and therapeutic armamentarium. The procedure consists of aspirating about 20 ccs of amniotic fluid from the uterus, at around the sixteenth week of gestation. The fluid and the cells that it contains are then subjected to different cytogenetic and biochemical tests. The list of genetic diseases that can be diagnosed by this method is increasing in length continuously . . . —*Extract from the paper by Dr Der Kaloustian.*

this kind? For one thing, the experts came from very different geographical and cultural areas, where the practice of terminating pregnancy has either now become accepted as a method of birth control (this is the case, for example, with certain East European countries); or may in some cases be encouraged in order to limit population growth (an example is developing countries with a high birth rate); or be tolerated as the lesser of two evils to prevent tragedies; or be prohibited by law, with the risk of a steep increase in the number of secretly performed abortions. For another thing, a discussion in a symposium of this kind would have been an admission that abortion comes primarily within the field of biology.[24] Now a decision to terminate a pregnancy may be based on highly varying motivations. Quite apart from abortions 'for the sake of appearances', and those which are akin to a form of birth control (which the author personally finds repugnant),[25] it will be recalled that the motives most frequently adduced for requesting a termination of pregnancy, in particular by young married women who already have several children, are strain and over-work, housing and working conditions, the feeling that they are unable to take on another (or a first) pregnancy, trouble with their childrens' schooling or with continuing their own education, their age (too young or too old), difficulties in the couple's interpersonal relationship, somatic illness, whether serious or otherwise, in which there is a risk that the mother's or the child's health may suffer, agonizing apprehension in cases where an earlier confinement has been difficult, too brief an interval between pregnancies, or a guilt complex in the case of young girls or adulterous wives. From this it emerges that the decision to terminate a pregnancy should be considered in relation to the child to be born and its prospects of integration in the family and in society. This being the case, even if it is usually the doctor who performs it, abortion is just as much the concern of sociologists, moralists and those in positions of authority, whether cultural, political or religious, as it is of biologists; and it is an evident scandal that many of the former should slough off their responsibilities on the latter.

The fact remains that, irrespective of the diagnosis and the circumstances under which action is taken, the doctor or the biologist has his own point of view to express with regard to what is called 'therapeutic' abortion. It is with this, and this alone, that we shall deal in the following paragraphs.[26]

A first observation is unavoidable on reading the extract just quoted from the Final Report (which, moreover, by its very vagueness reflects what tends to be the dominant attitude), namely that a shift has occurred in the concept of therapeutic abortion. Formerly this term was reserved for cases in which the life of the mother was objectively, and without any doubt, at risk if the pregnancy continued, particularly in the case of an extra-uterine pregnancy. It then became necessary to choose between two lives, so that medical practioners did not have the feeling that they were departing from the Hippocratic oath, since they were saving one life. In addition, they knew that in saving the mother they were also acting in the interests of the life of the father and the existing children, or those still to be born. Of the two evils, it was considered that the death of the foetus was the lesser, since the foetus was in fact in a less 'relational' situation. Today the term 'therapeutic abortion' extends to include the termination of pregnancy in the case of an affected foetus, though the life of neither the mother nor the embryo is threatened. How can we account for this change in view? There are those who condemn it utterly, considering it purely and simply as an instance of the loosening of morals. In the author's view, over and above the fact that it is linked to a whole range of psycho-social motivations reflecting the constraints of urban life, this changing view of the problem mirrors, on the one hand, and as a positive feature (though usually ineptly, to judge from the voluminous material produced on the subject), a deeper insight into the 'logic' and 'meaning' of the life of the child to be and of its mother, or of its family and social circle; and on the other hand, as a negative feature, a diminished sense of the tragic element in life.[27]

Let us first consider the 'positive' aspect. In discussions on abortion it is often pointed out that with animal species, a large number of embryos are eliminated whenever they suffer from serious anomalies; and this is used as an argument to the effect that, since human beings are less subject to such 'natural' selection, it is incumbent on us to remedy this by deliberately terminating a pregnancy when faced with the case of a seriously affected foetus. This argument should be relativized. Quite apart from the fact that on this subject we are the cultural victims of a kind of scientific infiltration which gives rise to a great number of morbid fears, it should be noted that, even in the animal kingdom, life has its 'failures'; in particular, we do not know what happens to all animals that are born: we know nothing whatsoever of the frustrations suffered by those with deficient 'powers of intelligence'; we merely note (lapsing naturally into anthropomorphic terms) that such and such an animal is 'stupid' or 'mad': perhaps there is a high proportion of animals with brain defects. But that which strikes us little or not at all with animals becomes of fundamental importance with human beings: for them, disease is characterized not only by its purely physiological effects, but also, above all, by the consequences it may entail for the relational character of the individual. And rightly so, since, as we have said more than once, this character is a prerequisite for the emergence of human life—of reason and freedom—and for its reproduction and its incorporation in society. Thus disease itself provides an insight into what human procreation is: not mere sexual generation, but a bringing into the world, into *this* world, *this* family, and through it, into *this* body of mankind which we at present constitute. Hence the crucial question leading to that of the possibility of a 'therapeutic' abortion: is society, and above all the family, now able genuinely to 'integrate' the child to be born? Would not the presence of a seriously affected child cause an upheaval in the home, polarizing all the mother's care at the expense of the other children and her husband, or creating ever-increasing material or relational difficulties which might even become unbearable?

The problem of abortion should therefore in our view be considered in relation to that which 'determinates' human existence, in other words its relational character, or to be still more accurate, in relation to a possible denial of freedom.[28] This means recognizing straight away that there is no *a priori* solution which can be applied generally to this problem, except a decision to respect as far as possible all the

different kinds of freedom involved. In this connection,
it is obviously not by chance that in recent years this
problem has been stated in so radical a fashion, at least
in some Western countries: just as musing on respect
for life led the ancients from the consideration of
murder and infanticide to an inquiry into the origins
of the individual, so it is only natural that, with a better
perception of the inherent dignity of woman (and the
relative alienation undergone by women in our society),
our contemporaries, posing the principle of respect for
the individual woman, her dignity and freedom, should
be led on to reflecting on maternity. Respect for
woman involves the whole spectrum of her fundamental rights—physical and mental health, the right to
love (there are circumstances in which the birth of a
child will inevitably break up the home), the right to
define one's own future freely (some unwelcome
pregnancies may cause enslavement, shatter a life
irretrievably or, for example, block a possibility of
marriage), and the right to exercise fundamental
responsibilities and to honour basic commitments
undertaken towards third parties (a case in point is a
woman's duties towards her existing children, which an
addition to the family might make it impossible for her
to fulfil). It will be seen that, in relation to the problem
of abortion, even therapeutic, the principle of respect
for woman goes beyond the sphere of her physical or
mental health, and encompasses what constitutes the
purpose of her life.

And wherein does this purpose lie *vis-a-vis* loss of
any *raison d'etre* (of the child, frequently rejected, or
of its family circle) or compared with deprivation of the
exercise of freedom? If a child is to be born which will
be relationally defective, or if lives are to be shattered?
In such a case, what is it that should be 'allowed to be'?
Absolute freedom can never *justify* the practice of
abortion: the termination of a pregnancy is always
both a self-disavowal on the part of freedom—which
expresses itself in 'allowing to be'—and the denial of
another's freedom. And nothing can change this. It is
solely and exclusively in terms of a real deprivation of
the exercise of freedom that the question of eventual
recourse to abortion can be posed. It may well be that
freedom should not under estimate its own capacity for
exercising itself, that is to say its ability to rise above

that which conditions it (thus it should not regard a situation as inextricable without due consideration, or be over-hasty in judging the harm which might result to it, or to another person, from the birth of a child). But if it is genuinely the case that no form of solidarity, even adoption, can loose the Gordian knot, it must be recognized that there is no predetermined solution to this problem. Freedom becomes aware of its contingent character, 'Free for what?' as well as 'Free from what?', no less and no more. It thus becomes, willy-nilly, an arbiter of life and death.[29] Hence the emphasis rightly laid in the Varna Final Report on the specific responsibility of the parents. The fact is not only that any outside pressure would be objectionable, but that this is an area in which it is impossible to draw up a catalogue of what is 'allowed' and what 'prohibited', and there is no generally valid reference to what has been done in other cases.[30]

It is nevertheless clear that freedom cannot bring itself to arbitrate in this way without careful and mature reflection on the decision to be taken. Here we have a question which was raised at Varna by Dr Der Kaloustian:

'Should we then perform abortions, even in the case of serious diseases, if in the future these diseases can be easily treatable? A good example is that of *'acrodermatitis enteropathica'*. This is a very serious disease, manifested by severe skin lesions and intractable diarrhea. Although the patients are mentally normal, they go through agonizing periods of suffering before they die, usually prior to the age of three years. A couple of years ago, if this disease were diagnosable *in utero*, we would have elected to abort the foetus. And yet, since last year, the disease is treatable in a very simple way, by the daily oral administration of zinc sulfate, resulting in complete recovery. In the face of such situations, our attitude, as physicians dealing with genetic counselling, has been to present to the parents the facts about the different diseases to the best of our present-day knowledge, and leave the decision to them. Under certain circumstances, and when hard pressed, we have expressed our personal opinion about our own possible moves, should we be facing similar situations ourselves.'

This is a very wise attitude, particularly since it

encourages the individual to assume his responsibilities in the light of motivations which go beyond what is strictly his own interest.

It is still true however that society can clarify or limit the exercise of this freedom, and if necessary lay down a number of strict conditions, specifying for example that any termination of pregnancy after sixteen weeks will be prohibited, bearing in mind various considerations relating either to the mother's health (physical, mental or spiritual) or its own specific requirements as a body constituted primarily to preserve the relationships between its members and to guarantee their survival in freedom. Here it should be noted that society cannot authorize abortion, since it involves a human life; all it can do is suspend judgement, recognizing that it has not the authority to take a decision in this conflict of freedom and conscious of its own contingent character and of its inability, of itself, to fulfil the purpose for which it is constituted. Similarly, it cannot in the last resort impart quality to the life of individuals, define the 'why' of their relational life or assign an ultimate aim to it; all it can do is provide the 'how' especially in view of the fact that it cannot define itself, since it is the result of these relationships.

Life once it has been given

The recognition that individual freedoms and society are contingent, in that they are living, is today more necessary than ever, especially in the West, since scientific progress has brought about an unrealistic confidence in man's powers together with gathering momentum in the headlong drive to improve man's condition. The result is a rejection of the tragic element, of the 'abnormal', sometimes of any kind of constraint, which as we have noted, explains at least in part the extended definition of 'therapeutic' abortion. But to limit oneself to considerations of well-being would be the equivalent of denying the quality of the individual, and would entail a regress of society, regarded as a mere aggragate of individuals grouped together in a field of opposing forces, without any real organicity. There is no doubt that we should combat disease, suffering and death, but not deny them, hide them from view or eliminate them from the fabric of society, thus running the risk of throwing off course the innermost forces which lead living beings to organize themselves in such a way as to assume joint responsibility not only for their self-preservation but for their procreation and 'destiny'.

For whether they like it or not, living beings are born conditioned individually and collectively by each other. To reject this conditioning, this 'destiny', is one of the major temptations of the world today, which chafes at its limits. But let us beware: any undertaking designed to force destiny can only engender another form of destiny, and whatever the ways in which life is manipulated, individuals will always have to accept one another. Suppose there were 'a brave new world': that one day 'test-tube babies' could be produced, or the physiological 'qualities' of the individual could be enhanced by a series of artificial devices; the children thus 'conceived' would not feel themselves any less determinate beings than we do. The rejection of 'destiny' sets mankind on an uneasy course of self-rejection, in which the individual cannot but feel desperately frustrated by comparison with his fellow beings or with the rising generation or future generations.

Without falling into the opposite error, which would be to consider that all is for the best in the best of all possible worlds, it is important to develop among men and women today a feeling of acceptance. Indeed this is the fundamental process whereby freedom manifests itself, in line with its most specifically original and 'originating' driving force: to consider as 'unacceptable' the narrowness of our existence, its flaws or faults, is an approach which is ultimately suicidal; whereas on the contrary, to accept them forces us to pay attention to the positive, and imparts a new impetus which often makes it possible to overcome constraints. We have only to think of those families in which a newborn child, often after an 'unwanted' pregnancy, is deeply cherished and brings untold relational 'wealth' to the home. Thus the acceptance of destiny is valid not only for oneself, but also for others. There is one way of denying the specificity of others, of trying to condition (if not plan) their existence, which proceeds from the rejection of destiny, perhaps even the rejection of 'otherness', and which leads to confining human beings within 'categories', reducing them to no more than a projection of oneself.

The above considerations are important, since today we are faced with two tendencies which have been strengthened by the progress of biology.

With the widespread use of contraception, we note a more marked propensity among parents to adopt a possessive attitude towards their child.[31] As parenthood falls increasingly in line with their desires and 'plans' (which may one day go so far as to include the choice of the child's sex), parents are apparently less able to tolerate any failure in the child to conform with the 'model' they have forged for themselves.

Concurrently, we observe a great increase in the number of children born out of wedlock, the result of copulation with a chance partner, and perhaps in the near future through recourse to a sperm bank. We have referred to the possible social repercussions should such a practice become widespread.[32] Today it is the case with young girls or women who, for a great number of reasons—disappointment, repugnance (due in particular to traumas caused by indulging at too early an age, or too lightly, in sexual relations), separation, loneliness, feminist demands, etc.—have no intention of linking their fate with that of a husband, and yet wish to have a child of their own, for themselves.

In both cases it will be seen that life is not really 'given'. No doubt a child is a projection of the parents' desires, an expression of their survival; it is rare for a child to be wanted for itself. Moreover it takes time a longer or shorter period, depending on whether one considers the father or the mother) for the parents to objectify the arrival of 'another'. Nevertheless, no child should be conceived without respect for this basic prerequisite of human life—the fact of being other, and being free. Whether we like it or not, we must recognize that all we ever do is to transmit a life which transcends us. We must also recognize that a child can only achieve self-fulfilment through the development of its relational character: for a mother to block that relation, restrict it to herself, is the equivalent of mutilating her child.

From both these points of view, we are bound to agree with the participants in the Varna symposium in categorically rejecting any eugenic project for selecting (or artificially reconstituting) gametes, fertilizing them, and then either transferring the being thus conceived into the body of a woman prepared to carry the pregnancy through to the end, or conducting the entire process from conception to independent viability *in vitro* (the test-tube baby). Even supposing that such forms of intervention are one day feasible (and we have noted that the second appears unrealistic), what would be the end sought? Would it be to create artificially a new strain of human beings? Here we come back to the objections already raised to genetic engineering: impoverishment, a loss of adaptability, concentration on present needs irrespective of future requirements, all jeopardizing the survival of the species. It may be argued that the aim is merely to produce a restricted number of individuals capable of performing a specific function which is necessary to society (e.g. the astronaut superman). To this our reply can only be even more conclusively negative, for such pre-determination would make the child the slave of a project, convert it into a social object, and limit its relational character along with its freedom. In a word, such practices would run radically counter to the 'logic' of the living being. It is true that evolution reflects in part the tentative search for an improvement in the human condition, both for society as a whole and for its members, for individuals and also for the species. But this search has led to the creation of fixed states: the fixation of species, incapable of evolving, or the equally rigidifying and anti-evolutionary fixation of individuals within animal societies (for example an ant-heap). The fact is that evolution also attests the emergence of forms of existence which are qualitatively higher. And fortunately so, since here again, there is no genuine or lasting improvement save in a driving force which produces such emergence, and it would be the worst possible mistake for mankind not to recognize this basic principle. And the task of raising individuals to a higher level of being is something that cannot possibly be pre-determined by their progenitors, any more than that of raising society to higher levels of being can be pre-determined by the present generation.[33]

Ethical requirements

To accept 'destiny' and otherness, to give life, to
remain open to the possibility of new developments . . .
all this involves risks. Yet it is surely our duty to reject
the hazards of destiny, to transmit 'liveable' life by
forestalling accidents (in particular accidents of reproduction) and to remedy variations from the normal or
cases of determinism. Here we touch on one of the
crucial tests of existence as such, which is ineluctably
caught up between the rejection and the acceptance of
risk. Biologists for their part are more and more
rigorously confronted with this problem as their powers
of intervention in regard to living beings increase.
And we would have no need of ethics were it not
necessary for us to discern the good, to impart quality
to the way in which individuals and society develop,
and to accept its problematical character as something
vitally concerning ourselves.

It goes without saying that the awareness of risk
subtended all the statements and discussions at the
Varna symposium, the central question being not so
much 'What good should we seek?' as 'What risks can
we take?'[34] It also goes without saying that this
question was not discussed for itself, or in the abstract;
a discussion of this type would in any case have come
up against a series of dead-ends: there is no discourse
which can give a final account of risk. If such a discourse existed, risk itself would no longer exist (only
choice would remain); furthermore, risk is life, and life
is risk. Thus one cannot ultimately 'justify' procreation.
To do so would be tantamount to not 'giving' life.
Has it not been said of parents that they are 'the great
adventurers of the world today'? This being the case,
we do not intend to expatiate on this topic either, or to
answer questions such as whether one is entitled to, or
ought to, take such and such a risk. Questions of this
kind are unanswerable, if posed in terms of the 'why';
all that can interest us is the 'how' the way in which
risk is integrated into the logic of the living being. A few
brief comments in this connection will give us a better
insight into the true situation as regards the innermost
driving force of life.

First, it will be seen that every living being spends its
life struggling with the risks inherent in its environent

or the special requirements of its constituent parts, righting imbalances as they occur and protecting itself against accidents and chance events. Admittedly, chance has in part determined evolutionary mutations; even so it has had to be accepted, made part of the organism, compatible with its programme and 'memory'. In other words, life does not integrate the haphazard except as a function of a 'certainty' (of its 'necessity').

Thus it appears that if we intervene in a living being, we have to modulate the risks in the light of what we know with certainty. The certainty may relate to the short- or long-term harmlessness of the manipulations or forms of treatment we undertake; or it may be the certainty that we can improve the organic equilibrium (and not merely lessen the effect of a particular disease). From this point of view, we would have to refuse to take a risk with a risk.

It is possible that this principle can guide us when we are faced with the hypothetical. It was on the basis of a situation of this kind that Professor Vartanian posed the following question at Varna: if it is true that some diseases are alleviated by treatment given before they break out, should not this treatment be started as soon as an individual comes into a high-risk category? If the treatment is undeniably beneficial, and its side-effects, if any, are at least preferable to the consequences of not administering it, there is little need to hesitate.[35] But what if the side-effects have serious and irreversible consequences, for example making the patient permanently dependent on others or impairing the genetic endowment? To revert to a case quoted by Professor Vartanian, should one give psychotropic drugs to all the children of schizophrenic parents, whereas only some of them (say, for convenience' sake, a third) have a predisposition such that they will actually contract the disease? There are doctors (though in point of fact very few) who advocate such preventive treatment: in such a case, two children out of three having schizophrenic parents would be abnormally affected, but for the third, the disease would be alleviated (in particular, this would obviate the need for hospitalization in a psychiatric institution). Faced with these two possibilities, Professor Vartanian stated: 'As a doctor, I have serious reservations with regard to measures of this kind, for two reasons: firstly, we do not know which of these children need treatment, and which run the risk of falling ill; secondly, the very principle of administering such a powerful drug to a child who is still in good health seems to me to be particularly questionable from the ethical point of view.'

This is indeed the case, since far from 'doing them good', this would be direct interference with the physical integrity of individuals who are in good health. And what about the others? How can one avoid feeling responsible for their fate? If the disease declares itself, will not the doctor be tortured by remorse, and the parents revolted, at the knowledge that it would have been possible, if not to avoid it, at least to alleviate its effects? But the prospect of making children in good health unnecessarily dependent on others is no less revolting, and moreover 'illogical'. One cannot regard what may be as more important than what is. The logic of life does not recognize the *a priori*, it judges and reacts solely in terms of acquired experience—though this does not mean that the living being does not organize itself in its own defence.

In this particular case, acquired experience is acquired experience only if assimilated into the organism. . . . We must consider the risks we take in terms of compatibility, regulation and assimilation, in regard both to individuals and to society as a whole. Thus we find equally 'illogical' a decision to institute a course of treatment regardless of its side-effects, and a certain tendency for medicine to become so specialized that general practitioners see nothing but the need to fight a particular disease, without considering their patient's present or future condition. Risks taken without reference to society as a whole are no less 'illogical': for example, without reference to its ability to underwrite the consequences of our efforts to help the individual (if only their cost); or to the individual's prospects of becoming integrated in society. In other words, the risk, above all the risk involved in procreation, must also be defined in the light of all that constitutes our solidarity, whatever makes us interdependent on each other, both negatively and positively. For all organisms, even if they are the result of chance circumstances and there has been no design attendant on their constitution, confer on each of their component parts both greater protection against random extraneous events and more adaptability to them. The whole contributes towards preserving the parts; and the parts, less under pressure from the need to defend themselves, are all the more free to make progress and develop new qualities in terms of their own specific functions. This applies no less to society as a whole. Since the human being lives in a state of solidarity or interdependence it becomes possible for him to take risks which would be fatal if he existed in isolation. This is evident from birth on, and even before. Admittedly parents should not conceive a child if they consider it impossible for them, and society as a whole, to assume joint responsibility for it, particularly if its health is seriously affected. Conversely, we should not be over-hasty in pronouncing an affected foetus to be 'impossible to live with'; to do so is to gainsay the solidarity that is one of the constituent functions and driving forces of society as a whole. In any case, would this not be to condemn society, as it were, to sclerosis? For while it is true that risk must be defined in relation to solidarity, solidarity finds its testing ground in risk, and draws strength from it. A risk, for the individual or for society, may open the door to new emergence, emergence to a new and higher level.

However, to say that one cannot take a risk with a risk, or that risk must be defined in relation to solidarity, brings up the entire problem of what we hold as certainties: both objective (to what extent will a particular cause produce a particular effect, or is a specific process set in motion, or is a specific intervention affecting one part organically compatible with the whole?); and subjective (to what extent can we count on the present and future solidarity of others, and of society as a whole?). Even supposing that we had full knowledge of the acquired experience and logic of the living being, it would still be true that with life there is no certainty, apart from the existence of a genetic code. Within the living being there is a combination of so many varied parameters bearing on its physiology and relational character that nothing can provide the key to what will come about. Already verifiable with unicellular beings, this takes on the nature of a principle with sexualization. To revert to the comparison sketched out at the beginning of this chapter, it should be noted that a bacterium reproduces itself by simple quantification (two from one) and, barring accidents, the duality has exactly the same kind of existence as the entity. Whereas sexualized reproduction subordinates quantification to quality (one from two) and makes survival dependent on a necessary reproductive difference, which also involves waste (waste of gametes, fertilized ova, etc.). There is thus no coming about, barring accidents or artificial means (e.g. cloning), save in the form of the identical lived differently: this need not necessarily involve a break in continuity (the genetic code and programme being transmitted from the original 'biosis'), but does at least involve rebirth. The sexualized living being is singular. Survival therefore ceases to be conceivable only in terms of physiology, and exists also in the relational, in a binary relation between two beings.

From these brief considerations it emerges that the risks we assume in biology should be modulated according to the individual's relational character, not solely according to his physiology. This also works negatively: one should not conceive a child if there are cogent reasons for considering that its ability to relate to others would be seriously impaired; nor should one pre-determine an individual genetically by aiming solely at improving specific physical abilities; nor should one embark on a course of preventive treatment (for example chemotherapy) which would unduly influence his mental make-up. Conversely, to enter into the logic of the living being—the sexualized living being—means encouraging and helping the individual to transcend himself, working ceaselessly so that he may equip himself with qualities corresponding to his singularity and his relational nature.

The paradox—the real paradox—is that, as sexualization vastly increases the risks involved (the meeting of two sexual partners, the reshuffling of their genetic programmes, all the 'adventures' of conception, fertilization, gestation, birth and early childhood), introducing into the living being a considerable degree of vulnerability, including structural vulnerability, by the same token it introduces (in a much more 'necessary' way than is the case with the duplication of cells) that which alone gives meaning to the living being, the true savour of life, even if life seems to be dependent on hazard and involves dramatic deviations: it introduces, in other words, the risk of new emergence.

It is in this light that we should view the immense amount of work put in by biologists.

'Even if the dangers to which current biological research gives rise are growing progressively greater, it does not follow that, yielding to fear or pessimism, we must aim to call a halt to such research. Rather, present-day discoveries should give us cause for rejoicing, when we consider the possibilities which they hold out for the progress of mankind. If we take refuge in pessimism, this means that we consider man to be incapable of recognizing and correcting his mistakes, and see him as the slave of his own victories, following a predetermined path. Such a presupposition runs counter to any ethic.[36]

It is true that we are to a great extent unable to gauge the risks we run, since many of our interventions will not take full effect until a future date. It is impossible to foresee the future. Having weighed the risks in the light of acquired knowledge, assimilation in the organism, solidarity, the relational dimensions, we must follow the 'logic' of the living being to the end and accept that sexualized life—introducing as it does qualities, the ability to relate, sublation—far from excluding, necessarily includes the risk of death.

Notes

1. Jacob, op. cit., p. 291
2. We elaborate this argument in the same way, attempting to bring out all its philosphical and religious implications, in *Cherchant qui Adorer*, Part I, Chapter 1, Paris, Gallimard, 1978.
3. Even if sexualization involves an enormous waste of living matter (it is well known, for example, that normally one human spermatozoid will fertilize a woman, whereas each ejaculation contains millions of spermatozoids; and that a minute number of young fish will be born from the thousands of eggs spawned by one fish, etc.).
4. See the chapter entitled 'The Order of Living Beings'.
5. We shall revert to this idea in the last chapter.
6. We expanded this argument in *Cherchant qui Adorer*, op. cit., Part III, Chapter 1. To say that sexuality in that event is devoid of morality does not mean that its exercise is necessarily immoral; simply it is seen to be amoral.
7. See above, pages 26–28.

8. i.e. thirteen out of the twenty-two days which is the normal gestation period for this animal. At this stage the heart, brain and limbs are already differentiated.

9. Without going into detailed scientific explanation, it can be said that after a certain stage, if the embryo is to continue to develop, it needs the mother's placenta, and that this placenta is so individualized that no human calculations can reconstitute it in all its special features. This at present makes any project to produce 'test-tube babies' unrealistic.

10. Today we are able to perform nuclear transplants, either replacing the nucleus of an ovule by that of another ovule taken from a female 'donor', or by transplanting into the ovule the nucleus of a cell, preferably intestinal, from the mother herself. Such transplants have already been performed on batrachians with successful results, i.e. the restoration of fertility.

11. This expression is typical of the attitude of those who are opposed to any form of abortion, on the grounds that from the moment of conception onwards there is a human being in the making. We have taken it from the title of an article by E. Pousset published in *Études*, November 1970.

12. It is pointed out in some quarters that the human embryo does not become 'individual' before three weeks, since up till then there is still the possibility of the emergence of 'monsters' (elements of organisms, constituted by fertilization taking place a few days afterwards, which 'graft themselves' on the already existing embryo).

13. The definition of life was not the subject of discussion at Varna: thus the reader should not regard the argument we advance here as representing a consensus subscribed to by all the participants in this symposium.

14. It should also be recalled that neither human intelligence nor human freedom can enter into play in the absence of a relation to another person. What would happen to our intelligence without language?

15. 'Phocomeles' are beings whose limbs are seriously atrophied. (Unpublished text by Professor Thibault.)

16. Professor Vartanian referred to this case at Varna. The analysis of the brain of the embryo (which is possible after a miscarriage or abortion) reveals that the tissues are impregnated by psychotropic substances.

17. See above, pages 73–75.

18. See above, pages 51–54.

19. This question was raised at Varna by Professor Vartanian.

20. Final Report.

21. ibid.

22. ibid.

23. Dr Mroueh devoted a paragraph to this practice as one specific birth-control technique. Dr Der Kaloustian made a brief reference to it in connection with therapeutic abortion, in a text which we shall quote further on.

24. In our view it would, however, have been valuable for gynaecologists to discuss the effects of repeated abortions on future pregnancies, (for example, the risk of premature babies) and on the incidence of physiological of psychological disorders which affect mothers in such cases, depending on the socio-cultural environment.

25. The reasons will appear further on.

26. Though some of the reflections put forward here can in fact be applied to all cases of termination of pregnancy.

27. It goes without saying that the following comments commit only their author and should in no way be taken as representative of the

standpoints adopted on this subject by the various experts who met in Varna.

28. Here we make use of the conclusion of an article we published in *Études*, October–November 1973, p. 407–23 and 571–83.

29. As it does in other tragic circumstances, such as wars and revolutions, or more immediately, in the choice of patients it is decided to treat and those who are left to die.

30. To draw up a list moreover might well encourage freedom to hide behind precedents, to evade what is in the matter at issue its specific responsibility.

31. This attitude has obviously many other causes, inherent in social life, particularly urban life. The individual, particularly the father, who is more accultured, feels himself more depersonalized and victimized, particularly by his working conditions; he wants his home to be a place in which he will find peace, harmony and the recognition of what he is: hence the need to lay down the law governing those immediately around him. To this should be added the desire for the child to 'succeed', that is to say, adopt the same image of success as his parents.

32. See above, pages 26–28.

33. We shall come back to this twofold 'tendency' of evolution in the last chapter.

34. See above, pages 29–31.

35. This is the case with vaccines.

36. Final Report.

Senescence and death

The good in relation to history

The prolongation of life and the quality of life

We have said that life, or rather sexualized life, necessarily includes death. The fact is that death is seen to be necessary for survival. It is necessary for the preservation and adaptability of the species, since over-population can bring about an impoverishment of resources which would create a risk of malnutrition, and also since the offspring of older individuals (who continue to reproduce in accordance with their own genotype) are more delicate and less adaptable than those of younger ones. It is also necessary for reproduction, since it is a fact, for which there is no real explanation (it seems that sex no longer acts as a stimulus), that overcrowded living beings tend to cease reproducing, and that sexual relations between progenitors in later life, when these are still possible, entail considerable risks of malformation of the embryo. Lastly, death is necessary for the purpose of allowing qualitatively higher forms of existence to emerge, since existence is largely dependent on the appearance and confirmation of what is 'new'. Regarded in this way,[1] especially in relation to the ability to evolve, death is beneficial to the species; the existence of the individual appears to be subordinated to that of the population as a whole and of future generations, on condition that he imparts to his life a quality which will enable him the better to give life, that he accepts, in other words, the necessity of self-sublation. It accordingly seems as if death enters into the genetic programme itself. As Professor Bezrukov said at Varna:[2]

'The expectation of life is a characteristic of the species. The fact points to a link between ageing and changes in the genetic apparatus of the cell. There are two distinct theories about this: (a) ageing is a result of genetic programming and consists of quantitative and qualitative changes in the genetic apparatus which develop in time following a regular pattern; (b) ageing is a result of damage and lesions which occur in the genetic apparatus during the life of the organism and are the results of accumulated errors in the DNA molecule . . . Those who hold the former view consider that ageing, like morphogenesis and embryogenesis which are controlled by genes, is governed by a "biological clock". The proponents of the second theory explain ageing in terms either of a single cause, such as spontaneous mutations, free radical reactions or crossing over, or of multiple causes including the influence of a large number of metabolites, physical and chemical changes in the protoplasm. Ageing seems to be a complex process associated both with programmed changes and with the action of metabolites.'

Reviewing these various changes undergone by the organism, Professor Bezrukov showed that each of them is both effect and cause, and that there is 'no single cause or single mechanism' which explains 'the highly complex and fundamental process' of ageing.

Can we, however, be content with the theory that the living being experiences a period of growth up to the point at which it is capable of self-reproduction, and thereafter, having accomplished its work on behalf of the species, embarks on a slow process of degenerescence? Even physiologically speaking, things are not so simple. To quote Professor Bezrukov: 'It has become clear, that, paradoxically, besides the destruction and degradation of tissues, ageing entails the emergence of important adaptive mechanisms Longevity is largely contingent on how strong these mechanisms are and how well they work . . .

'Such adaptive mechanisms are, of course, far from perfect, and even if they function as they should they do not suffice to halt ageing. They help maintain the homeostasis of the ageing organism but the regulatory systems nevertheless become less reliable, the capacity for adapting to the environment declines and susceptibility to disease gradually increases . . .

'Ageing is not a return to earlier stages in the life history of the organism, to earlier levels of metabolism and functions and of their regulation. Ageing is a natural stage in the ontogeny characterized by a qualitatively different relationship between the various levels in the system of control of the metabolism and functions.'

It would seem then as if, albeit doomed to disappear, the living being was nevertheless subject to an all-pervading process which slows down and regulates its own decline; as if the organism tended to grow old 'as a whole', in a balanced way, without any abrupt break in any function; as if 'life' abhorred formlessness and continued to respect the 'principle' of organicity even in the ageing process. These observations are borne out by the fact that one of the main changes which are then seen to occur is a kind of reorientation of the neuro-humoral regulation system. Here again we shall quote Professor Bezrukov's words:

'We regard changes in the neurohumoral regulation system as highly important, firstly because they have an important role in maintaining a particular level of metabolism and functions and to some extent prevent a breakdown in the activity of various organs and systems in the ageing organism, and secondly because they limit the possible range of reactions of old people and determine the specific nature of their adaptation to changing environmental conditions.'

Thus recapturing the 'logic' of the living being implies acceptance of this process of slowing down and regulating decline. What is the situation as regards present possibilities for increasing longevity? Obviously, infant mortality must be reduced. While considerable progress has been made on these lines in the developed countries, where it is in the region of 3 per cent,[3] much remains to be done in the developing countries, where it is still as high as 6.5, 10 and even 20 per cent.[4] It is also common knowledge that there is a need for work to continue on the provision of medical facilities and the improvement of living conditions throughout the world, since the average expectation of life remains less than 30 years in a number of very underprivileged countries in the continent of Africa, and countries with unhealthy climates in South America and Asia, whereas it reaches 67.5 years for men, and 75 for women, in industrialized countries. In latter countries, the average expectation of life would be increased very little by the eradication of diseases and accidents.*

* The eradication of diseases of the digestive organs would increase the average expectation of life in the U.S.S.R. by 0.13 to 0.22 years, and that of tuberculosis of the lungs and of other respiratory organs by between 0.16 and 0.49 years. The greatest gain would come from the eradication of malignant growths (an increase of between 3.03 and 3.53 years) and of injuries and accidents (a potential increase of 1.16 years for women and 3.5 years for men).—*Extract from the paper by Professor Bezrukov. It should be noted that these figures correspond approximately with the estimates for a country such as France, except in the case of accidents and violent death, where the increase would be only 0.89 years for women and 1.94 years for men.*

The situation would be quite different if one day we applied to human beings 'a number of methods . . . already worked out whereby the expectation of life of laboratory animals can be increased by 20 to 30 per cent or even more.'[5]

These methods can be either tactical or strategic. 'Tactical' methods are those designed to realize the potential expectation of life of the biological species: for this purpose human beings for example would necessarily have to adopt a reasonable and regular mode of life, a balanced diet, a certain amount of physical exercise and a suitable organization of their work. 'Strategic' methods are aimed at slowing down the rate of ageing, and are concerned with the feasibility of altering the expectation of life of the species itself (in particular by a better control of the neurohumoral regulation system). However, any such strategy would be pointless unless backed up by tactics. Hence these lines from the Final Report:

'For the individual, preventive and protective measures, mainly nutritional, hygienic and educational, are urgently needed from the time of birth, or even from conception, in the hope of averting much of the intellectual and physical deterioration of old age. Such considerations have a much higher priority than research on the prolongation of life to 150 or 200 years.

'All participants, scientific and philosophical, were of the opinion that the aim should not only be prolongation of life as such but, more specifically, active longevity—physically, morally and psychologically— thus avoiding many of the ethical problems that society has to face with old people today . . .

'Further studies should determine more accurately how far the period of life for man is genetically predetermined, and whether the full use of a man's faculties can be carried up to such limit, before any major attempt is made to prolong life further, as the ultimate aim should be to "add life to years, rather than years to life". This striking phrase epitomizes more clearly than anything the common opinion of scientists and moralists.'

Here much remains to be done. The developed countries have been taken unawares by the increase in longevity and by the number of old people. In the U.S.S.R., for example, the average male expectation of

life was 31 years in 1896/97, 42 in 1926/27 and 65 in 1970;
old people over 65 represented 6.8 per cent of the population in 1939 and 11.8 per cent in 1970.[6] In other
countries, chiefly European ones, the percentage of old
people is higher still. There then arises, over and above
the problem of the cost of this population, the problem
of the conflict between generations, particularly in
periods of economic recession, when young people see
their road barred, if not to employment, at least to
positions of responsibility, since many elderly people are
reluctant to give up working for fear of not having
enough to keep them going or feeling cut off and rejected. This is another example transposed to the level of
society, of the 'vital' tension referred to earlier between
younger and older individuals within each species.

It is then tempting to say that, since such tension is
ineluctable, growing old is a depressing fate, which
condemns the individual to a gradual rejection for the
'world of the living'. Such an attitude would in fact
seem 'logical' if survival were defined solely in terms of
reproduction. But to repeat what we have already said
above: first, with sexualization, reproduction is relativized: while the entire existence of the bacterium is
devoted to constituting on its own two bacteria, the
sexualized living being does not reproduce on its own,
and does not spend all its life in self-reproduction. The
procreator of new life, the mediator for a possible new
emergence, it is a being in which the relational character
develops; its entire evolution thus seems to be in the
direction of a greater opening up towards others, in
relation to whom, and through whom, the living being
'acquires qualities'. Secondly, the new forms of regulation which take shape during the ageing process are also
seen to be essentially of a 'qualitative' nature, designed
to preserve not only organic equilibrium but also (we
venture to say so by inference from internal forms of
regulation and from the adaptation to the environment)
the living being's integration with its surroundings and
the maintenance of its relational character.

These comments give food for thought on the place
of the ageing of the individual in contemporary society,
and indeed on what constitutes life in society as such.
It may be asked whether social life does not concentrate
too much on the requirements of conservation and
reproduction (in fact, production), at the expense of

the acquisition of qualities and new emergence. We cannot here answer this question, to which we shall revert at the end of this chapter; but for the purpose of what follows, it will be useful to bear in mind already at this stage that it is a question posed by the 'logic' of the living being, and that in the last analysis it relates to what is meant by the survival of individuals and of society as a whole.

The approach to death

How astonishing are the processes which enter into action in senescence, slowing it down and regulating it! They are valuable to us as signs. Contemporary biology has achieved remarkable progress as regards resuscitation and the prolongation of life. But what kind of life? And offering what kind of assimilation in society? Take for example the treatment of hydrocephalus. Today we know how to remedy it, by means of some fifteen surgical operations over five years. A triumph of medicine? Only 10 per cent of children so treated will have an approximately normal conscious and relational life; the others (whose compressed brains have been unable to develop along with the rest of their bodies, or have suffered lesions) will be idiots . . . What fate is there in store for these young invalids and their families?[7]

The fight against disease, even when successful, does not necessarily imply a cure, or a return to a more or less normal life. As we have seen,[8] the trend today, at least in some hospital centres, is to concentrate on reducing disease (through types of treatment which are becoming increasingly 'heavy'); while in some cases forgetting about the individual patient. Now if attention is concentrated entirely on action to combat disease, rather than on a cure, the decision to undertake a particular treatment is no longer the same; it becomes more 'automatic' and more impersonal; discernment is involved to a lesser degree; and the responsibility is less onerous, in that the battle is waged against an adversary, a specific disease, which is moreover considered in the abstract, as one case among many others of the same kind.

Let us not hold practitioners to blame for this. The

progress of biology and medicine faces them daily with increasingly heavy responsibilities, with agonizing questions to which there are no ready-made answers, since the very problems themselves have changed with advances in techniques and the revolution in outlooks. Some of these questions were brought up at Varna.[9] We shall try to set them out for the reader, not so much with the intention of finding a general solution as of throwing light on what, in our view, should be the approach to the decisions to be taken.

Respecting the integrity of life

The Hippocratic oath admittedly enjoins on the doctor the duty of doing all in his power to save human life. But what is meant by human life? It is to be put on the credit side of medicine that it is possible for men, women and children to lead an existence which is entirely dependent on artificial devices, without any lasting or deep effect on their intelligence, affectivity or ability to relate to others. If in such cases there is still the problem of physical or moral suffering, this should not cast doubts on the doctor's skill. But what is the situation when conscious and relational life appears to be confined to vague, confused glimmerings, sometimes intermittent, against a background of total darkness? Or an even more obvious case, when the individual is reduced to a purely animal, not to say vegetable existence? Throughout this study, we have sufficiently insisted on the essential character of relationality to venture to say that, in our view, any being once irreversibly deprived of the ability to relate to others in any way, and denied self-awareness, can no longer be regarded as a human being.

But in the present state of medical knowledge, we do not know exactly what conscious life is, or even, except in case of widespread destruction of certain brain centres, at what point the loss of consciousness has become irreversible. There are situations which are relatively clear cut, where there is unfortunately no ground for subjective hope. Sometimes we know that a process has set in which leads inexorably to the loss of conscious life . . . But are we ever sure of the duration of this process, or that the patient will not live through more or less lengthy 'stationary stages', before continuing to go slowly downhill into darkness? There are cases of wretched beings, tragically handicapped in some of their mental faculties, for example hemiplegics, who nevertheless retain astonishing sensitivity and clear-headedness, sometimes even real creativity. On the other hand, how many of those affected by the same illness, and in a similar state, become impatient, vindictive, lachrymose, making life very hard for those around them ... We have thus a whole rising or falling gamut of states which it is impossible to predict in advance. In such cases it is the doctor's strict duty, based on respect for the patient who has put himself in his hands, to be cautious in giving a decision, and to fight to preserve life so long as there subsists an objective possibility of conscious life.

Conversely, respect for the patient can also lead to discontinuing treatment, refusing to continue the struggle for survival when there is no further possibility of recovering a certain degree of consciousness. The common expressions used about someone who has fainted, that he 'comes to', or 'is himself again', give food for thought. What is meant by 'himself'? Not merely a physiological organism. Is he still 'himself' if he is a being who admittedly breathes, but whose relationship with others is irretrievably disrupted, who is permanently unrecognizable to others and for ever incapable of recollecting this world, or being present in it? The life we seek to prolong, with its disfigured features—is it not in fact an abstract entity, or a social taboo, or a spurious 'good'? And can the fear of serious abuses be enough in itself to give one an entirely free hand, and justify conduct which, at a certain stage, no longer relates to the individual as such, but to his shadow, or to his trace in society? In other words, when we place our trust in a doctor, we do not on that account give him permission to tend what would be a substitute for us, merely a distorted image of our true selves, nor to dissociate the conscious from the physio-logical life within us.

Similarly, is it not distorting death to consider it solely from the physiological point of view? Even if we confine ourselves to this aspect alone, it is a fact that death engulfs us by degrees, in a process which may be more or less rapid, starting with the loss of con-sciousness and ending with the termination of all bodily functions and the decomposition of the tissues. It is thus not possible for us to perceive death as such.[10] Even supposing the patient genuinely recognizes that he is 'at death's door'—and there are many who deliberately delude themselves, who think they are only 'dangerously ill', and who fluctuate between hope and renunciation—it remains a fact that the dying man falls into a syncope before drawing his last breath; and this will probably be increasingly often the case.[11] Many patients nevertheless feel that the end is near, or rather that they are at the threshold of death. But for the human being, the passing of this threshold cannot be reduced to the single test (however painful) of the interruption or cessation of his bodily functions. There are probably very few who make of it the supreme act (as preached by many philosophers) which confirms the purpose of their existence: the last cry in the face of destiny, the last challenge, the last acceptance or recognition of their condition, or the last act of faith and hope that there is 'another' life. At least we must recognize that death is the referent in relation to which we feel the dramatic, contingent, temporal character of our existence; that the fear of death, even if kept out of sight, haunts us continually; and that we are instinct with a death wish, as revealed by depth psychology. This being the case, and whether we are ready for death or taken by it unawares, death cannot be dissociated from our existence. In other words, dying is always relative to life, this life: the lives of those around us, or our own. If this is so, it is obvious that we all retain a certain 'right' to ask to be allowed to die, consciously to accept our own passing. And it may possibly be the worst of all insults to deny death to a human being by wrongly and artificially retarding it, or by completely exhausting all the vital forces. Yet today, on the pretext of combating death, not only do we 'depersonalize' it, we also make it 'inhuman'. This should not be taken solely to mean that, in fact, we impose on the patient a period of survival which is sometimes accompanied by unbearable suffering; we also seek to smooth over the passage to death; to treat every instance of death as something which comes from the outside, as an accident, as a technical failure or inadequacy; briefly, to deny the human being his own natural death, and the possibility of accepting it, however it may come about. Yet if, knowing himself to be free yet also mortal, the individual is caught up in an irreversible process of losing consciousness, has he not the right to refuse treatment which would alienate him in this ultimate approach to himself, or else would merely prolong an existence in which he would be irretrievably cut off from effectively exercising his freedom and his ability to relate? In our opinion, this right is theoretically inalienable.

We should, however, beware of the ambiguity of the word 'right', used in connection with respect for life and the approach to death. While today there is an increasingly vocal, and justified, demand for the patient and his family to have the right to participate in vital decisions—the beginning or discontinuance of a specific kind of treatment—it would be dangerous to enshrine this demand in some unspecified legal formula. To adopt such a position would be the equivalent of endorsing precisely what is objected to: on the one hand, the considering of medicine as purely a question of techniques, and the exercise of medicine as merely the compulsory choice of the best technique, or in other words, the standardization of forms of treatment; on the other hand, 'standardization' of the patient and the illness. It should be reiterated that the course of an illness is individual, it affects a single case, and it is treated by an individual doctor or an individual medical team, in hospital conditions which vary from one country to another, one establishment to another and even one hospital department to another.[12] This does not mean that all that needs to be done is purely and simply to relativize the exercise of medicine;[13] it is a recognition of the fact that, irrespective of any question of competence, medical ethics in the last resort is based, and can only be based, on the trust the patient has in his doctor, and the respect shown by the doctor for his patient. But here again, is it respecting the patient to submit him to any or every kind of treatment? And does the trust we place in the doctor, as responsible for our person, mean that we leave our fate entirely in his hands? In any case, can we hand over all responsibility for ourselves to someone else, when what is involved for us is a matter of life or death?

Accepting death

The patient's 'right' to have the integrity of his person respected; his 'right' to be respected in his approach to death . . . We have here two claims to entitlement, which are moreover overlapping, and cannot be reduced to any clear-cut system, which are in our view basic, and which it is apparently agreed should be honoured by society and the medical profession even if they postulate a refusal to be treated, and if it is known that sooner or later death will be the result. However, it should also be borne in mind that the individual's prerogative of refusing is not such that society does not have to ratify it, or define how it shall be exercised, in the same way as society is entitled, for example (as we have seen above), though the analogy is somewhat remote, to limit the exercise of another basic right, that of parenthood. The fact is that society as a whole cannot fail to be involved in any decision to refuse treatment, at three levels. First, out of respect for the individual, and for the common good and the good of his family. Second, just as radically, because society is constitutionally involved whenever the fate of one of its members is put in question. Third, because, no matter how much the existence of an individual is a personal matter, and no matter how irreducible his freedom, at no stage of his existence is he able to realize himself, or give expression to his freedom, without other human beings; these are as necessary a 'condition' of his being as is his own body; he cannot therefore isolate himself from them, even in his approach to death.

Conversely, while it goes without saying that no one can assume the right to eliminate an individual deliberately, the case may arise where society as a whole has to honour and ensure respect for every patient's basic 'right' to be respected for himself, if it is impossible for him ever 'to be himself again' or to lead a conscious and relational life.

What is meant by 'society as a whole'? In this particular case, it naturally means all the individual's family, his near relations; generally speaking, all the other individuals in the society, and first and foremost, on account of their functions (and the task delegated to them by society), the members of the medical profession. In practice, the bringing to fruition of a

decision to speed up the process of dying is a matter
for the individual, the family and the doctor; but no
single one of these three parties can isolate itself from
the other two, nor from the body of society as a whole.

These considerations may appear over-theoretical;
but they have very specific implications, which should
not be underestimated. Let us consider the state of the
question as regards participation in a decision to refuse
treatment by each of the three parties most directly
concerned: the individual, the family and the doctor.

The individual. We have just called for respect for the
patient and for the dying. Does not such respect consist
in the first place in informing the individual that his
days are numbered? Many people hold this view, and
it is common practice in some countries, for example
France, many doctors are opposed to 'preparing' their
patients for death. Such opposition has its reasons,
which should not all be dismissed as an abdication of
responsibility. We shall confine ourselves to two of
them.

As we have already said, attitudes in the face of
death vary considerably depending on the individual:
some display resignation and courage; others refuse
to accept it, and demonstrate their will to live up to the
end, sometimes even aggressively; others refuse to face
it, and keep up an illusion; others regress to a state of
near-childishness, letting themselves be fussed over by
nurses as if they were babies; some literally die of fear.
There are in fact as many ways of approaching death as
there are individuals, and it is by no means sure that
the man who asserts his will to live up to the very last
instant, even hoping against hope, and without any
illusions, shows a less noble or free spirit than the one
who goes clearsightedly to meet his death. If we
consider it lacking in respect to deceive the patient as to
his condition, the same is true of imposing on him a
'pattern for dying'. At the very most all we can do is to
accompany him on the way (to the very slight extent
that this is possible), thus answering the appeal he makes
to us, whether it relates to trivial realities or is expressed
in requirements of a spiritual nature, whether it be to
help him fight on to the end or not to prolong his life
unduly, or to make the conditions in which he
approaches death more human.

There is a second important reason for the reluctance of some doctors to tell the patient the truth: the desire not to cut him off from those around him, not to give him the feeling of being rejected. And indeed, how many patients (particularly those suffering from cancer), once informed of their condition, give up the struggle? And a great many patients (in particular children), while aware they are going to die, act out a part with those around them (as in turn do those around them).[14] Here we come back to the importance of the relational character: the being who no longer relates 'dies' before 'perishing', before life actually departs. It is often said of a dying man that he is 'condemned'. The common use of this expression is a sign of a very grave inadequacy of contemporary industrialized and urban civilization (we shall come back to it at the end of this chapter): in an entirely consumption—and production—oriented society, there are no longer any words for approach to death: all is silence and rejection, death is 'de-socialized'.

These considerations show that no one can judge in advance what will be his own reactions at the approach to death. Since the subject was brought up briefly during discussion at the Varna symposium, we venture to express our view that one cannot regard as decisive the 'last wishes' expressed by someone in good health about the way in which he sees his dying moments. Moreover one must in our opinion be on one's guard with this kind of 'testament in advance', in which healthy individuals ask (or would ask, if it were recognized practice) not to be resuscitated or treated in specific circumstances, or if they are struck down by a specific disease. Where such a request is drawn up by an individual at a time when he considers illness in the abstract, is it still valid when his health has been impaired, and in view of the fact that, as we have already said, illness is always an individual case? What is 'cancer' in the abstract, how does it strike each of its victims, how does it evolve, hence at what stage does a testament in advance become valid? Furthermore, progress may be made which, in the space of a few weeks, alters the purpose of the request by, for example, enormously increasing the chances of a cure. In addition, stress should be laid on the frightening 'snowball effect' which would be the result of general introduction of such a practice. Anyone signing a 'paper' requesting that, for example, in the event of hemiplegia, he should not be given treatment, so as to avoid costing too much to society, or not to be a burden to those around him, or for any other reason, quite apart from the fact that he does not know what exactly would be his conscious state of existence in such an eventuality, would exert undue influence on his relations: should not they also adopt the same 'good intentions', and draw up a similar request? And should they not also act likewise in the event of other diseases, such as cancer? Little by little, should we not reach the stage of considering that if someone has not drawn up such a 'last wish' it is because he has not thought about it, or did not have the time? But for those who know how unselfish he is . . . and had he really never talked about it . . . ? One can see the excesses to which this might lead. In reality, the major flaw in this kind of 'testament in advance' is, in our view, that it relates primarily to the fact that man necessarily 'perishes', and that it entirely overlooks what cannot be 'felt in advance'—namely one's reciprocal relationship with others in the approach to death, it prejudges the fact that one is 'condemned', and endorses 'de-socialization.'[15]

If we reject such action in advance . . . are there any patients who can be recognized as sufficiently free to take a decision which involves their existence, at a time when present, for them, is riddled with suffering, fear weariness or loneliness? Particularly since there are very few who are capable of diagnosing with certainty the seriousness of their condition, or the irreversible nature of their illness. This being the case, an appeal by a dying patient not to have his life prolonged unduly, or to have the conditions in which he approaches death made more human, even if we recognize that there is an obligation on us to respect it, must necessarily be looked at critically, precisely on account of the trust and respect a dying patient is entitled to expect in his relations with others. Who, then, is to take a decision? His family?

The family. The responsibility of the near relations is in fact involved in any decision to adopt or discontinue a type of treatment. This is true for many reasons, which

it is unnecessary to go into here. It is obviously the
case when the patient, still conscious and relatively free,
can express his wishes: the family responsibility is
involved in backing him up or, as the case may be,
dissuading him in view of other considerations. It is
also the case when the patient is unconscious, either
having lost consciousness permanently, or where
he has never acquired it (the case of babies with severe
prenatal or perinatal traumatisms), or where he is too
young or ignorant to express his wishes.

This is a responsibility which devolves individually
on each family, not only because it concerns one of its
members, but also because the family forms a kind of
physical or legal entity which is relatively independent,
and because it has its individual features and requirements. For example, can one impose on a family,
without due reflection, the burden of either a gravely
handicapped child or a very aged person who has
'degenerated'? This is an unanswerable question when
posed in advance of the circumstances, since many
elements enter into play, which can only be known in
each case. The most commonplace (housing situation,
working hours, etc.) are not the least decisive; but
account must also be taken of the psychological
make-up of each member of the family—for example,
it should be considered whether a particular mother
will not transfer all her affection to the handicapped
baby (probably with a latent guilt complex), at the
expense of her other children and her husband. The
presence in a family circle of, to take one example, a
child or old person with a serious brain disease may be
a factor making for exceptional maturity, or be the
cause of tragic disturbances or loss of stability. Here
again, it is impossible for the outsider to decide. It
remains none the less true that it may be an intolerable
abuse of the family's credulity, or a betrayal of the
trust placed in him, for a doctor to begin treatment
without informing at least the patient's relations of the
effects which can reasonably be expected therefrom, and
of the state in which the patient will survive. If the
patients has a right not to have his life prolonged as a
mere travesty, his relations should see that this right is
respected; just as in the same case they have their own
right (which they exercise, or may exercise, in one of
several ways, for example by sending the patient to a

specialized establishment when the burden of care is too heavy for them—it is not for us to judge whether they are being selfish) to refuse the prolongation of 'a' life which would no longer be that of a member of their family.

However, there are many doctors who are reluctant or even extremely loath to involve the responsibility of the family when it is a question of beginning or continuing a type of treatment. Is the family capable of participating in such a decision? Even supposing that all its members are unanimous, or that there has been a valid decision taken previously (by whom? generally speaking, or in the light of a special case?), or that one particular member can sway all the others (will he not be a target for criticism from all those closely involved?), countless other difficulties arise. Some are due to the fact that the relations themselves experience suffering, fear, weariness and even a feeling of being cut off, in the presence of the illness which has struck down a member of the family, and that they too, like the patient, are rarely capable (sometimes even less so) of making an enlightened and realistic diagnosis: are they, therefore, really free and consciously aware of the issues? In addition, people's temperaments vary (sometimes temperaments run in families), as do cultural and social environments. The result is that there are as many different approaches to the death of a loved one as there are families, or even members of each family. How are we to generalize, and can the doctor adopt a standard form of conduct, or even suggest a 'model'?

It should moreover be noted that there can be no extempore participation in such a serious decision. The doctor, for his part, knows roughly what will be the evolution of the illness (he has a certain amount of 'statistical' insight into it); he has also been faced with a great number of similarly dramatic cases which have come to a tragic end, in spite of him and no matter how agonizing his own grief, without any reflection whatsoever on the way he has discharged his responsibility. To the parents, for example, of the child who suffers and is being tortured, it may well appear inhuman to prolong life; and the father and mother are prepared to listen to a doctor and rely on his advice, particularly since the more deeply distressed they are the greater his influence over them. But if he decides to let the child die, after a few hours or days or weeks a feeling of guilt may insidiously set in, which may or may not be recognized and shared by both parents: Was there nothing that could have been done? Shouldn't they have consulted another doctor? Didn't I give in too soon? Wasn't I deceived, or wrongly persuaded, by my husband (wife)? And so on. Conversely, what parents are capable of appreciating the physical or moral suffering likely to be entailed for their child or the family by a mounting series of medical interventions which may possibly lead only to a seriously handicapped form of survival or, worse still, a near-animal existence? In such cases it is understandable for a doctor to refuse to take the risk of traumatizing a family; he is respecting the relational considerations as much as the sociological ones.

The doctor. For his part, the doctor has his own three-fold responsibility. He must respect the medical team of which he is a member, without overlooking any of the eventual human implications of the decision to be taken, either for individuals or for the team as such. There is no doubt, to take an elementary example, that to perform a tracheotomy to save the life of a patient who has already been the exclusive concern of several nurses for days and nights on end, may constitute an unbearable burden. And will not the additional strain and attention involved be reflected in a lowering of the quality of care given to other patients? The doctor is thus led to choose, from among several patients, those to whom he will apply, or will refuse, 'heavy' treatment (i.e. treatment which is labour-intensive or requires special apparatus), in cases where there is a shortage of staff or equipment . . . Lastly, he must clearly assess critically, from his own professional standpoint, any request by the patient or his faimly for treatment to be begun or discontinued.

But is the doctor really in a position to assume his particular share of responsibility? If he suspends treatment, what does he know about the possible reactions of his team? And on the pretext of not imposing too great a strain on the team, does he not run the risk of traumatizing one or more of his assistants, or giving them a guilt complex? If he has to 'choose' between two patients, by what criteria will he decide? And is he even in a position to give the assessment expected of him as to the appropriateness of treatment, when, as we have seen, the evolution of a disease is always more or less individual?

To sum up, whether it be the patient, the family or the doctor, it is clear from the foregoing that no one is fully equipped to take a decision of such grave import.[16] In that case, should one avoid taking a decision, and leave it to technology to decide? Or could it in fact be the case that the patient loses his individuality, and that when it is not possible to do so, there is no need to respect either his desire not to survive only as a semblance of himself, or the way in which he wishes to approach death? The answer is most obviously no.

But the fact remains that there is no 'model' for this kind of decision, any more than there is a standard patient, only one approach to death or only one reaction to the death agony of a loved one. This being the case, we must repeat that the doctor is the person who is best placed to take a decision. It may be that he does not seek to take over the responsibility from the patient or the family, if they are incapable of assuming it. But to be loath to seek it is by no means the equivalent of rejecting it. And the fact of having to assume it on his own does not mean, to put it briefly, that the decision is his, nor, *a fortiori*, that he is entitled, any more than anyone else, to modulate it as he sees fit. Here there arises a crucial question for biologists today: To what extent is one entitled to hasten the moment of death?

Redefining death

'Nature' is not the only arbiter of death: what man can do, or is powerless to do, also enters into play, if only to modify the way in which this 'fatal issue' comes about. More and more frequently, except in the case of sudden death (which is rarely 'natural', but more often takes the form of accidents, including accidents in the course of treatment, or homicide), death takes place when man gives up, whether it be the patient himself, his family, the medical team in charge, or society as a whole—where, for example in a developing

K

country, society refuses to provide itself with modern hospital equipment, preferring to invest in fields other than that of health. This 'giving up' may be variously motivated (in order to shorten suffering, not to monopolize an apparatus, etc.), due to circumstances which are also extremely varied, and it may take very different forms, such as discontinuing treatment or refusing to discontinue it. Death thus becomes more and more relativized.

The distinctions which even up till recently provided a basis for moral judgements with regard to euthanasia are therefore tending to disappear. It is thus generally recognized that it is permissible to give sedatives to a patient to attenuate his suffering, even if taking them will hasten on his death.

'If the dying man consents, it is permissible to make use of narcotics in moderation in order to alleviate his suffering, though they also bring about death more rapidly; the reason is that in this case, death is not directly intentional, but is inevitable, and well-balanced motives justify measures which will hasten its coming.[17]

This text sums up fairly well a view shared by most of the experts who attended the Varna meeting. But in practice, it raises a host of questions, such as: What is meant by the consent of the individual patient? By speeding up the process of death? By 'well-balanced motives'? Above all, what is meant by 'in moderation'? The fact is that the means available to us for alleviating suffering are becoming increasingly powerful (ranging from chemotherapy to surgical operations to 'disconnect' certain nerve centres).[18]

Similarly, it was recognized that there was no obligation to take extraordinary measures to prolong the life of a dying man.[19] But what is 'extraordinary' in one case may not be so in another.[20] The expression 'leaving it to nature' is equally ambiguous: is it still possible to refer to a 'natural' order, when our entire existence is permeated by the artificial and the civilization to which we belong, whether we like it or not, is inherently interventionist in character, imbued with the idea that the human being is not entitled to evade basic responsibilities, either on his own behalf or on behalf of all mankind? Who is now to draw the line between the 'natural' and the artificial? Even the former distinction between 'active' euthanasia (bringing about death) and 'passive' euthanasia (allowing to die) is difficult to maintain: for to cease blood transfusions or disconnect artificial respiration is in fact a deliberate act—an act of relative powerlessness—designed to bring about death. We are thus increasingly faced with a new responsibility, that of giving the possibility of dying. When, and in what conditions, can one assume such a responsibility?

Professor Malek recalled at Varna:

'An absolutely safe definition of death continues to be one of the most difficult problems in medicine . . . As a rule, the currently used criteria of cerebral death—profound coma, areflexia, respiration arrest, flat EEG—are sufficient. Yet occasionally, they do not permit of making unequivocal conclusions. Therefore, additional appropriate criteria are sought that would, for example, in cerebral angiography, ensure a more accurate and rapid diagnosis.'

The search for such criteria for establishing death is necessary for the 'protection and reassurance of the public.[21] It is of great importance in cases of organ transplantation. Such transplants particularly kidney transplants, are bound to increase considerably, for a number of reasons: because techniques will gradually be developed (this is already the case with kidney transplants); because in many cases (for example, liver or heart transplants) this is the only way of prolonging a patient's life; and because in other cases it is the least expensive method (this is true of renal transplants, since the need to have frequent—or even, less frequent—recourse to dialysis involves much greater cost, and is also more tiring for the patient; a patient receiving artificial kidney treatment costs, in France, some 100,000 francs a year, and lives for about ten years, and each year there are 2,000 new patients to be added to the list for such treatment).

It is clear that one cannot remove a vital organ which is unique, such as the liver or the heart, before the death of the donor. But even in the case of double organs such as kidneys, the tendency today is to remove organs preferably from cadavers: this at least appears to be the case from Professor Malek's paper: 'The present trend is marked by gradual departure from the use of kidneys from living donors owing to the ever-improving results of cadaver kidney programmes. It

relieves the physician of having to decide whether or
not to recommend donation. The donor incurs a double
risk. In the first place, there is the risk of the actual
nephrectomy which, despite being minor, must not be
under-estimated—for example a case of death during
this operation was reported in France [Professor
Grosnier]. In the second place, the donor has but one
kidney [left]. The risk involved is virtually nil, as the
compensatory hypertrophy develops in the remaining
kidney. On the other hand, the surgeon is faced with
the dilemma of having to perform an operation not
benefiting the patient. There is a world-wide, though
not an even, tendency to abandon the use of kidneys
from living donors. Some centres, for example in the
United States, continue to transplant kidneys mainly
from living donors. In Czechoslovakia, the consensus
is to accept kidneys from living donors in exceptional
cases only, i.e. siblings with identical HL-A systems[22]
and in response to the donor's insistence. However,
even in such cases, severe criteria are employed. Great
care is taken to establish beyond doubt the validity of
the prospective donor's consent. It can happen that
consent is given under moral pressure from another
member of the family, and the donor himself has some
doubts. We have found this to be often the case with
adult siblings in relation to their own family. Despite a
25 per cent probability of identical HL-A systems, the
probability of a conflict of interest of the 'old' and 'new'
family cannot be discounted.'

But the removal of an organ from a cadaver must
be done without losing any time: the earlier the better.
Hence the importance of defining as strictly as possible
the moment of death. There may be a great temptation
to decide too quickly, in the interest of the recipient,
that a potential donor is dead. In order to prevent
possible abuses various systems of regulations have
been drawn up in different countries: one regulation
which affords the maximum guarantees is the obligation
to have recourse to two separate medical teams, the one
performing the transplant being distinct from the one
responsible for removal.[23]

The foregoing considerations relate to cases of
'sudden' death, usually accidents. But contemporary
medicine is also faced with the converse problem,
that of patients who, maintained with artificial support,

'go on dying indefinitely'. Individuals in a state of prolonged coma may be kept alive for years on end, even when there is no conceivable chance of recovering consciousness, monopolizing both medical staff and apparatus which might be used for other less seriously affected cases. The question therefore arises whether we are justified in putting an end to such artificial survival.

From this point of view, it appears essential to define not only physiological death, but also 'human' death. As we have said, in most cases the human being enters on death by stages, in a process of decline which has no fixed time limits; but from the moment that he finally loses all possibility of conscious, hence relational life, he is bereft of an essential characteristic, and we should consider what is the medical significance of maintaining an existence which will never see the individual 'be himself again'. This may be justifiable in the interests of science, but certainly not as regards respect for the individual. The author's view is that, when it can be said of an individual who has entered on the process of dying that 'no one can do anything more for him'—no one, as a person, being in a position to do anything concrete to overcome this helplessness in relation to him as a person—that individual has then lost an essential condition for the exercise of his reason and freedom. In our view, such an individual is humanly dead.

Even if the statement of such a view were recognized in principle, it would still raise numerous difficulties. First, of a medical kind. In the present state of our knowledge, it is in fact difficult to make a sure diagnosis: there is no intellectual coefficient which is seen to be decisive; and in some cases, the irreversibility of the process which has set in is open to doubt. (Have there not been cases of patients whose EEG is apparently flat, even recorded over several days or weeks, who recover a certain degree of consciousness?) It is to be hoped that scientific research in this field will be continued more actively, so as to achieve either greater accuracy in the reading and interpreting of EEGs, or a more rapid spread of new techniques for determining to what extent brain life still exists. This is assuming that we have full knowledge of all the physiological conditions governing conscious and relational life.

Other difficulties would arise, in redefining 'human'

death as we have suggested above, from the socio-
cultural environment. For it must be recognized that if
we attach great importance to defining death solely in
physiological terms, it is because the criteria appear to
be more reliable, or because we hold a religious respect
for life; but it is also, and probably still more so,
because we dislike the idea of our fate being dependent
on others (and though we do not realize that a medical
definition of death based on purely physiological
findings may be questionable). From the standpoint to
which we have referred, a medical 'certificate' of human
death is just as much relative to the attendant circum-
stance of the helplessness of the family circle as to the
rigorous character of the diagnosis; thus it does not
proceed from principles which can be made universally
applicable, but is a matter of relativity (or rather
relationality); its morality can only be judged in relation
to the honesty of the assessment which plays the
decisive part.

Giving the possibility of dying

In point of fact, these proposals apply only to patients
who have entered on an irreversible process of dying,
whose death is predictable within a fixed period, and
who have lost both consciousness, and with it their
relational ability. We still have the very great number
of cases in which the individual remains more or less
conscious, though stricken by a fatal illness which has
only a short course to run. In this case we consider it
justified to discontinue treatment at the request of the
patient, even if it means bringing about death, provided
that such a decision is duly examined by the family
and society, as suggested above. It is true that death is
a concept related to life. But the converse is not true:
human life cannot be defined as the absence of death;
it is qualified by the significance we impart to it. This
being the case, what would be the significance of an
intervention supposedly designed to 'help to live'
whose only result would be to 'prevent from dying',
to prolong the absence of death? What would be the
significance of treatment, or a surgical operation, which
would do no more than delay death or prolong the
death agony, other than that physiological life is

sufficient to define the individual, regardless of what gives meaning and purpose to his existence? Thus society cannot oppose such a desire on the part of the patient, except if he is mistaken about his condition (about the incurable nature of his illness, or the proximity of death), or if the common good is directly involved. Society should show a certain amount of 'humility': not only is it not entitled to impose a significance on individual existence, since it would thus alienate the freedom of its members, but it does not itself constitute the sole meaning and purpose of the individual's existence. Conversely, it is generally agreed (and reiterated in the Final Report of the Varna symposium) that no one can refuse treatment to a patient requesting it (save in totally unjustifiable cases, or where it is totally beyond his power—for example if the cost is prohibitive for the family or society), even if there is apparently no chance left of the treatment being effective.

There are, however, all too many cases in which the patient cannot take the decision, because he is either too young or not in a condition to make a clear or considered judgement. We have seen that the family or the doctor could take the decision which in their view shows the greatest respect for the patient whom they are responsible for 'helping to live' (rather than 'preventing from dying'). Is one entitled, systematically, to refuse to acquiesce in death? Such an attitude may express a denial of the mortality of the individual, a kind of desperate wishful thinking; or it may show a lack of responsibility, or blindness, which when all is said and done is lacking in respect for the dying patient. In reality, a systematic refusal to acquiesce in death usually reflects the community obsession referred to above, or not leaving the fate of an individual in the hands of his fellow-beings. This obsession has two aspects. A negative one, of fear: we mistrust others. such only by virtue of his relationship with others, with society as a whole, society must recognize that its mission is to 'help to live'; thus we repeat that society cannot generate death, without disavowing itself as being made up of individuals and as representing the organicity of their mutual interrelationships.

Nevertheless, if society has a bounden duty to help to live, and it is by this means and for this end that it is in fact constituted as society, it should also be noted that we are talking of 'human' life, which is relational and looks to emergence on to a higher level of being. This obligation of society *vis-a-vis* the individual never ceases throughout his existence, right up to his death. But it has its limits, in respect of the individual, of the finite nature of the human condition (in particular, the unreliability of communication) and of society as a whole, which is itself finite in the way it is constituted, and frequently powerless to provide the conditions which are necessary if existence is to have meaning, or if it is to be possible to perceive this meaning. To recognize this powerlessness, and to admit that society, or the family circle, does not represent the ultimate meaning of free individual life, or the elixir by means of which it can emerge to a higher level of being, is not to be lacking in respect for the individual; it is at all events the realization that, for the human being, death is not merely something governed by fatality, or an external cause (a simple fact of 'nature'), but forms part of the relational character of the human being; abandonment by others, the loss of relationship, the loss of meaning are no less causes of 'human' death than is an impairment of the psychosomatic conditions of existence.

To put it briefly, in the event of a fatal illness where the end is in sight, provided there is no opposition by the patient (assumed, or taken for granted), it may be permissible to give the possibility of dying, that is to say to discontinue all forms of treatment, even to disconnect an apparatus prolonging life artificially, and deliberately to assume the responsibility of the resulting death. But such a decision can only be taken in the very last resort, when we are finally forced to recognize that all else is in vain.

Refusing to give death

Giving the possibility of dying, allowing someone to die, is not the same as giving, or putting to death. Do we have the right, in order to put an end to a terminal illness in which existence has become 'insensate', to bring about a physical process of death other than that which has already set in: for example, to administer poison, in one way or another, to a patient who is in terrible pain and who asks for his agony to be terminated? The question is now posed openly, backed up by apparently humanitarian arguments and motivated to a great extent by the relativization of death in the world today. Our reply to it, in common with all the participants in the Varna symposium, inasmuch as they adopted the Final Report, is in the negative. We should, however, adduce our reasons.

If one takes only the individual, there is no obvious reason why, if he asks to be put to death, declaring that he does not wish to prolong an existence which is henceforth devoid of purpose, his request should not be respected: as we have said: What right have we to impose a pattern of dying on him? But the individual cannot be taken in isolation. In our view, quite apart from other philosophical or religious considerations, the rejection of euthanasia[24] is linked to the relational character of the individual and is based, to put it briefly, on respect for the common good and for society as a whole. As will be clear from a few practical considerations, and brief reflection of a more theoretical nature, to admit that one is entitled to put to death would mean destruction of the most elementary social relationships. And this is true, even without stressing here the risks of error or abuse: mistakes in diagnosis as to the irreversibility of the process of dying which has set in, and its evolution, or an extension of the idea to cover a state in which death is admittedly drawing near but is not foreseeable (as for example with the very aged, who are definitively bed-ridden), or as the desire to get rid of a relative who has become a burden, etc.

Who would take the decision to put to death? We have seen that even a decision not to undertake treatment poses serious questions; giving or putting to death would raise still more alarming ones. Is it for the patient to decide? But what exactly does he know about his condition, and if his suffering is acute, is he not likely to react unthinkingly? His request should be subject to review by several doctors (in order to obviate a mistaken diagnosis). In addition, such a request should be made in due form (in the presence of impartial witnesses) in order to prevent any suspicion falling on the medical team or on the family circle. The procedure should be the same for a request made by the family in the belief that it is acting in the interests of a patient who is not in a condition to decide. But which member of the family would be empowered to take such a decision? Should we therefore recognize that the medical team alone would decide, in all good faith and in secret? This is to assume that the medical team would be unswayed by emotion, and entirely unimpressionable; and secrecy would be a disaster, since it might lead to doubts as to whether any and every case of death in hospital has not been artificially brought about. Lastly, who will be made responsible (and by what forms of pressure) for putting the decision into effect? And how will those responsible for the decision and putting it into effect be regarded in their family and professional circles, and how will they judge themselves, at the time or later on?

It will be argued that, if euthanasia were recognized by society, these objections would gradually disappear. But one might just as easily assume the converse, that custom would engender new risks of abuse. In reality, a society which legalized the practice of euthanasia would be inciting the two groups most directly concerned in helping a patient to live—i.e. his family and the medical profession—to act in a manner contrary to the 'logic' of life, in the sense we have used this word throughout this work; in so doing it would be fatally poisoning the springs of its own lifeblood as a social entity.

But, it will be argued, surely the same applies when it comes to allowing someone to die, giving them the possibility of dying? Not so. And most of us feel this obscurely, even if we do not manage to formulate the reason for it.[25] The reason may perhaps be found in the distinction we have already briefly drawn above: depending on which of the two conditions necessary for the exercise of freedom are destroyed (that enabling him to invest himself with new qualities and that giving him access to a higher level of being), the human being will perish when he loses 'physiological' life, and will die when he loses 'relational' life. To give the possibility of dying to a patient whose existence has no longer any purpose is to accept the fact that man will always 'perish' to recognize that the human being is not purely a biological organism to be kept alive at all costs; it is not directly infringing in his relational character.

Conversely, to give or put to death is in the first place to sever the ties of 'relationality', and only secondarily to destroy the physiological conditions of existence.

Possibly a clearer insight may be gained into what we are trying to show by taking the example of suicide. There are two kinds of suicide. There is the 'noble' suicide by those who prefer to die rather than to betray their deepest commitments;[26] and the 'obscure' suicide of those driven to desperation. Surely it is clear that, in the first case, suicide is a testimony of the respect felt for certain social obligations, or for that which gives meaning to life. The suicide knows that, if denied this particular relationship with society, he would never be the same person, that self-sacrifice is better than severing the ties which bind him to society or, *a fortiori* grievously damaging it. 'Obscure' suicide is also an expression of the fact that the individual no longer feels at home in society (we are not concerned here with ascribing responsibility). It signifies in the first place the loss of the relational character: something has 'died' in the individual concerned so that he feels shut out from society, and all that remains for him (as a necessity, or a last cry) is to perish.

Euthanasia is often portrayed as a kind of 'noble' suicide, with the individual who feels his life being engulfed by the meaninglessness of death preferring to put an end to it rather than go into a decline or be a burden to others. But this is to consider only one aspect of the problem, the purely individual point of view, without reference, on the one hand, to the common good (since being a burden on others, far from destroying society, may be one way of constituting it as such) and, on the other hand, to the place we occupy in society (since our life is not our own individual property, and its purpose must always be considered in relation to the way it forms part of that society as a whole). Whether he likes it or not, the individual does not represent pure, untrammelled freedom, but freedom embodied in society. And what we say here about the individual is also true of his immediate circle: pity for the sufferer, and humanitarian reasons, should be considered from the dual standpoint of the individual and society. Humanitarianism cannot be left to the discretion of individuals (since the individual is not humanity): it is a matter for mutual consent, for society

as a whole; indeed, it is in the constitution of society
that it is first manifested.

In reality, euthanasia is from the factual point of
view much closer to an 'obscure' suicide (whether it be
the individual himself who takes his own life, or one of
his family who takes it for him), suffering having made
the relationship with others unbearable. And it may
well be that there is nothing we can say or write about
this which will not appear entirely impertinent. For it
would mean assuming that society was itself a perfect
entity, with the tragic element no longer having any
place in existence, and afforded a full, meaningful and
liberating relationship with others, even in the worst
cases of suffering. But just as we feel our finitude in face
of the physiological conditions governing life, so our
relations with others are basically affected thereby and,
generally speaking, suffering isolates people rather than
binding them together.

To sum up, therefore, it is the author's view that
society should reject euthanasia, that it cannot possibly
endorse it, since it is totally unacceptable to it, but that
it should abstain from judging it as if it were an offence.
(This does not mean that we should not probe the
circumstances of a suspicious death, in order to curb
abuses.) We should probably seek to devise other forms
of legislation or regulations, other types of legal
machinery to enable society as a whole to assert its
responsibility and requirements in keeping with its
mission of 'helping to live'. There is a limit to justice,
as there is a limit to freedom; but there is no kind of
'logic' which can impel society to forego its own survival, any more than that of each of its members.

Ethical requirements

In concluding this chapter, and confining ourselves
solely to what is strictly the 'logic' of the living being,
it is clear that the 'phenomenon' of death can be viewed
from many different standpoints.

Physiologically and individually speaking, death
indicates a kind of genetic de-programming, a breakdown in organization, whereas taken in relation to the
species and society, it is seen to be genetically programmed, organic, bound up with renewal, and com-

prised in a relational dimension. From this point of view, death is not only an autonomous process affecting a particular individual being, but a law governing (and helping to constitute) a particular species; the 'ceasing to be' of the individual becomes the means by which the species—or, the society—can subsist and rise to higher levels of being.

But death's scattered shafts can only usher in the preservation of the species, and its emergence on to new levels, if the essence of that which disappears is reproduced: otherwise the species—or the society— would disappear in turn. The forces deployed by the living being, the 'programme' which governed its existence, is stored in the 'memory' of the species, and the function of the individual within society must go on, even after his death. In other words, there is never any 'ceasing to be' in the 'logic' of life.[27]

From this point of view, the following remarks are called for.

It is worth reflecting (as we suggested in connection with senescence) on the 'logic' of a society which does not 'partake' in death, but tends to exclude or 'desocialize' it, thus showing its disregard for one of the basic ingredients of the lifeblood which imbues it. It is true that this attitude prevails in Western countries; other cultures or religions, African or Asian, are much better at integrating death. This is no the place to analyse the causes of this 'desocialization', which began as far back as in Graeco-Roman and Christian antiquity,[28] and which has recently been accentuated under the influence of the so-called 'consumer society'. The fact remains that (to keep strictly to our subject) by proscribing death, by portraying it as a kind of decay, avoiding talking about it, or trying to exclude it, we have on the one hand made it a subject of still greater anxiety, the cause of countless neuroses, with repercussions for the psychological and physicological equilibrium of individuals; and on the other hand we have indirectly encouraged practitioners to show excessive zeal in embarking on quite absurd courses of treatment. A thorough change in outlooks will have to be brought about—and we might pave the way for this by enlightening our contemporaries on the biological necessity of death—so that generations to come may be able to face up to this new responsibility which we have seen will increasingly devolve upon them, that of giving the possibility of dying.

However, in order to assume such responsibility, it is also necessary that each individual should be aware of, and consciously recognize, the fact that not only is he indebted, if not for his existence, at least for the opportunity of deploying his own qualities, to the death of those who have gone before him, but also that he is imbued with the identical life-blood which imbued them—and which he must transmit in his turn, being also doomed to die. Only he who thereby 'integrates' his own death, and prepares for it, can allow someone else to die in the full awareness of what he is doing. For such a decision would be 'inhuman', if the individual concerned did not take it in relation to his own death. In other words, allowing someone to die pre-supposes that each individual feels himself involved in death, as being a fellow-subject of death, because he is similarly involved in life, even though, like life, death is always unique. Here we come back to what we suggested earlier: death should be 'integrated' not simply as the act of 'perishing' but as the act of 'dying', that is to say as occurring in the context of a relationship with others.

If this were the case, we might well feel a lessening of our deep aversion at the idea of our fate being dependent on others. For when one really thinks about it, it is no more offensive to our moral sense to allow someone to die than it is to cheat him of his death by forcing him to continue in a state of 'non-dying'; and we may perhaps one day come to the conclusion when reflecting on the seriousness of bringing into life, that it is no less honourable for man not to die as the result of an accident or chance circumstances, but to receive his death at the loving hands of those to whom he has given life. Death would then lose some of its hatefulness.

And surely this is to be desired. For the horror death inspires in us hides this from us; yet death (biologically programmed and integrated), perhaps even more than sexuality, reveals the paradox of life: we have said it before, and we repeat, that an overall view is necessary, that death does not exist in the abstract—only living beings exist—but nevertheless operates not only in the context of a binary relation-ship—between living beings—but also 'beyond the frontier' of the living: in other words, death shows itself to be essentially historical. This aspect, whose importance we have glimpsed on several occasions (in particular, in connection with the problems of genetics and procreation), is crucial to any attempt to throw light on the relations between biology and ethics. Let us take a brief look at it.

What in fact is life, if not that which 'happens' in and between living beings ? The living being is a moment, a locus, an instantiation of life. This is proved over and over again: this instantiation is always finite, and less than perfect: no living being can achieve to perfection what is postulated by the 'logic' of life. It is for the biologist to keep these deviations within bounds, to counteract the illogical (particularly when it is likely to bring about an 'un-programmed' death); but he would be seriously mistaken if he tried to determine existence on a uniform pattern, preventing life from adventuring itself or trying itself out in each individual. On the contrary, he should do all in his power to give each such adventuring every chance of success, and see to it that the conditions are fulfilled which are necessary for it to continue. When one thinks of the countless hosts of those who once lived and are now dead and gone, it may seem that such adventuring is vanity. Yet there is no form of life (if only that which provides food for another) which can be dismissed as being without importance: both because it is that which enables life to continue, and because it contains the potential of new emergence. What would make non-sense of the existence of an individual is not the fact that it will come to an end: it would be for life to end with itself, if life did not open out on to any future, if it became 'dehistoricized', if no living being took over and carried on the lifeblood which inbued previous generations. If this is the case, we repeat that biology cannot be the science of 'non-dying', but the science whereby life transcends itself through bringing into play all its latent potentialities.

We thus see that while the living being is the subject of his life (admittedly, to be exact, we can only say so of the conscious subject), and must strive to impart new qualities to it, to continue the adventure it represents, he nevertheless does not in himself hold the key to its meaning; this can only transpire in and through subsequent generations, to the extent that they preserve, genetically, at least the basic physiological features of life in the form adventured by their forebears. Putting it briefly, we must again repeat the truism that, biologically speaking, life is memory and expectation, turned towards what-is-to-come.

It is from this dual point of view that biology also acquires meaning and value. It is incumbent to biology to recapture the past, the innermost necessities, the programme or memory of each life, so as to enable the living being to continue with its own adventure, and also to preserve and prepare what-is-to-come. We should not be alarmed at the idea of biology being adventurous, or opening out on to the unknown. What would be terrifying, indeed insensate, would be on the contrary for biology to crystallize living beings in the state in which they exist, to standardize them on a predetermined model, arresting history by confining life to nothing more nor less than reproduction of the identical.

There is no history without at least potential new emergence. As we have repeatedly reminded the reader, one of the major prerequisites of such emergence —and also of the 'direction' taken by evolution—is seen to be the grouping of living beings into increasingly structured and regulated organisms, and the growing complexity of the relations between individuals. As the science of what-is-to-come of life, biology must be the science of the conditions governing these groupings and this intensification of relations.

But can biology turn in exclusively on itself? What, in fact is that which is to come? Where is history leading? What, if any, is its meaning or purpose?

Notes

1. We could have added other comments on the same lines.
2. His paper was signed jointly by Professor V. V. Frolkis (who did not attend the meeting).
3. These are mainly cases of perinatal mortality or death from congenital anomalies.
4. Infectious and parasitic diseases are still very widespread in these countries.
5. From Professor Bezrukov's paper.
6. Figures taken from Professor Bezrukov's paper.
7. In view of the tragedies that have been occasioned as a result (the anguish of families, followed by bitter disappointment which may cause a catastrophe in the home—not to mention what the child

feels, or does not feel, since it is impossible to perceive it), the specialist who first developed this form of treatment stated, in an article which was given wide publicity, that for his part he was discontinuing it.

8. See above, pages 107–110.

9. Dr G. E. W. Wolstenholme and the author spoke on these questions. No lengthy discussion was possible for lack of time. It is therefore difficult to say whether a 'consensus' emerged among the participants on the substance of our statements, the Final Report remains somewhat vague in this respect, particularly since those who spoke did not approach the subject from the same standpoint. We thus feel bound to repeat here, with even greater emphasis than before, that the following pages commit only their author (the gist of this argument was published in the November 1974 and December 1975 issues of *Études*).

10. We shall see further on that one of the problems posed for biologists today is the definition of death.

11. Bearing in mind, for example, the power of the analgesics administered and the techniques developed to overcome suffering, which have a very profound effect on the nerve centres, also bearing in mind the fact that death is increasingly retarded, and often takes place only once the patient is already severely 'failing'.

12. Many other considerations might be gone into here to show that illness cannot be regulated by 'law'. This would carry us too far from our subject.

13. In any case, in most countries doctors have their charter, and various professional or governmental bodies ensure its application.

14. Many patients speak of their death to their nurse rather than to the family or the doctor (who is more remote and more representative of society).

15. These comments should not be taken as applying in any way to permission given, even long before death, to remove an organ from one's own body. Some countries, such as the United States of America and Canada, have given official recognition to this practice, which has the merit of avoiding family conflicts in the case of a removal for the purpose of a transplant. In point of fact, such conflicts are tending to die out: Professor Crosnier reports that 'four years ago, 85 per cent of families in France declined permission for the removal of an organ. At present, permission is obtained from 90 per cent of families'. (Extract from Professor Malek's paper.)

16. Even so, we have not referred to certain aspects or situations which make it more difficult to find a 'ready-made' solution.

17. From a speech by Pope Pius XII, on 9 September 1958, repeating what he had already said in an earlier speech, on 24 February 1957, Dr Wolstenholme referred to the principle thus formulated, which has been widely endorsed by the medical profession.

18. It should also be noted that surgical techniques have made great advances, particularly from the point of view of their 'finish', so that operations today have fewer physically painful after-effects than in the past.

19. A principle also enunciated by Pius XII (see note 17 above, and his speech on 24 November 1957).

20. At present, the only criterion for distinguishing between what is 'ordinary' and what is 'extraordinary' is cost. However, cost is itself relative, depending on the establishment (the town and the State), and on whether a type of treatment is becoming common practice or not (for example, an artificial kidney does not cost the same today as fifteen years ago).

21. Final Report.

22. The expression 'HL-A system' designates a chromosome sector which is 'responsible for tissue compatibility' and governs a number of cell 'markers'. Each individual possesses six markers for each cell, three inherited from each of his parents. However, these six markers are themselves 'chosen' hereditarily from among a great number of 'varieties' (some fifty of them have so far been identified and analysed). (Explanatory note by the author.)
23. This is the system adopted in Czechoslovakia (as also in other countries), the advantages of which were demonstrated by Professor Malek.
24. In the passage which follows we therefore take euthenasia as a synonym for 'bringing about a process of death other than that which has already set in, in particular in order to shorten suffering or the death agony'.
25. Since we are not able to do so, we adduce a number of prohibitions, generally on religious grounds.
26. For example the suicide of a combatant who has been taken prisoner, of a person in danger of being raped, or the acceptance of martyrdom for a cause.
27. One example is the progressive constitution of the embryo of a mammal, which, as is well known, briefly passes through all forms of life, starting with purely cellular life (it even goes through a branchial stage).
28. Cases in point are, on the one hand, the Platonic concept of the soul as it is commonly understood, namely a spark of the divine enclosed in base and shamefully perishable flesh; and, on the other hand, a certain interpretation of Christianity which presents death as the 'fruit' of sin, in a theological system which focuses on redemption. Here we refer the reader to our essay *Cherchant qui Adorer*, op. cit., Part II.

Life—a phenomenon beyond our mastery

The good in relation to ex-sistence

Where is life leading? What, if any, is its meaning or purpose? In the preceding chapters we set these questions aside and focused our attention on the more immediate problems faced by biologists. However, they cannot be evaded if we wish to clarify the relationship between biology and ethics, and take a critical view of scientific advance.

In our view, a relevant ethics (the 'functional' ethics suggested in the Final Report on the Varna meeting)[1] would be derived from biology and imply an understanding of the bio-'logical' order and the dynamics of life. But what do we know about living beings? We can categorize their physico-chemical constituents. We have succeeded in mapping out the genetic code. We know that genetic information is passed from parent to offspring by means of a programme which will be relatively simple with the least 'organic' forms of life but increasingly complex as one moves in the direction of greater organicity. Supposing we had full knowledge of this genetic programme and were in a position to modify it, what would be the outcome of such an advance? We would obviously be able to counter disease and other disorders. We would be better equipped to defend ourselves, individually and collectively, in the relentless struggle against a whole range of external dangers and possibly even greater dangers from within ourselves. We would, perhaps, one day be able to take the chance element out of genetic structuring. We could deploy biological advances to achieve a quantitative and qualitative improvement in production and provide against the pressing needs brought about by the inordinate rate at which man is today consuming vital raw materials. But is it enough simply to ward off the dangers which assail us? Is it enough simply to satisfy our most urgent needs? If we were to stop there, we would be dealing with effects, not their cause. Surely biologists can legitimately set their sights higher than this? Should it be their aim then to impugn the forces which blindly govern our fate, to gain real mastery of life? Yet, supposing man were to achieve such aims, where would that leave him? What advantage will he derive? Seeking emergence into qualitatively higher forms of existence, how is he to define them?

These are not questions for consideration at some future time. They concern us here and now, even if our apparent aim is merely to relieve certain deficiencies or combat various dangers. What is our objective? And *how* shall we attain it? Evolution cannot be hurried along in an arbitrary fashion, but we must at least grasp what has made it possible in the past.

Life: an unending process and an ultimate design

We do not for a moment suppose that we can resolve the controversy among scientists regarding this question.[2] We merely wish to put forward a number of points which in our view form an essential background to the development of an ethical standpoint.

We shall begin by briefly stating the opposing views (even at the risk of oversimplification). In view of the closely structured and regulated nature of biological systems, it would seem inevitable that their genetic programme, if it is not to disappear, must simply reproduce itself. It is impossible to escape from this internal necessity. Accordingly, if mutations occur, they will bring about a modification of the actual genetic information, without any loss of fundamental compatibility. On the other hand, mutations are necessarily random since, as we have already stated, self-reproduction is an essential feature of all genetic material. The random product is acted upon by selection on the basis of its individual and environmental conditions. Thus, to take an example we have already referred to, out of millions of microbes, a single organism, with a unique constitution, will survive and proliferate in the face of antibiotics. The evolution of the microbial species is not to be attributed to the antibiotic (except in rare combinations of fortuitous circumstances). It is due, rather, to the prior existence of a particular genetic trait. Natural selection acts only on an *a posteriori* basis; it does not create anything. Admittedly, recent Japanese research, especially the work done in this field by Motoo Kimura, has shown that variations at bio-molecular level are neutral and are fixed quite randomly. It would, therefore, seem that the concept of necessity (at least at this level) is less rigorous than might at first appear. In actual fact, a very large number of mutations affect bacterial populations. However, the Neo-Darwinian is not interested in establishing why a mutation occurs. He is concerned with how it is integrated into the genetic material and becomes organic. 'Compatibility' is the key idea here. The question of an ultimate design does not enter into consideration.

Nevertheless, to explain evolution as the outcome of random mutations selected on the basis of necessity or confirmed as it were by chance (provided the random product is compatible with the existing organism) is no explanation at all. The Neo-Darwinians have not yet produced an answer to the classic objection to their theories, namely that the gradual formation of the eye, for example, would presuppose thousands of millions of mutations which are sequential and organic (if only in regard to the nervous system). In each generation, these mutations would occur and be selected in very small numbers, if not on an individual basis. And each generation would have the task of transmitting the 'new trait' to its offspring, who in turn would have to continue the process. Furthermore, all this would have to happen at random. If the 'project' of the eye were not genetically predetermined but executed at random, countless billions of generations would be needed before the eye could emerge (which is not the case) and selection would have to proceed unerringly towards the goal (which is incomprehensible since it can only be discerned how beneficial the eye is to the living being once the organ has at least roughly formed). Sexualiza-tion is an equally bewildering phenomenon, despite the fact that it has its archetypes even in bacteria (certain of which are, as we have already mentioned, more capable of 'receiving' new information, while others have a greater disposition to 'transmit' such information). In particular, the mutations which induce the emergence and differentiation of the sexes must necessarily operate in an ordered and synchronous manner, producing different affects in the two lines of what we cannot yet call descendants, so as to obtain, in the one, males and in the other, females. In view of this, it is difficult to understand how anyone can fail to accept the purposive character of the 'phenomenon' of life.

However, if this is the case, at what point does the predetermination first occur? The position of the 'hormicists' is no less shaky than that of their opponents. If the genetic programme has remained essentially unchanged since the first living beings, does this mean that the eye and sexual differentiation were implied in the first assemblage of organic matter which began to function and constituted the first living cell(s)? Why, then, has the evolutionary process involved so much 'groping', so much 'trial and error'? Why is there such a plurality of forms? Even though life may be purposive, everything happens as if it were obstructed, at least in some measure, in the pursuit of its ultimate purpose, while chance alone seems to release it by slow degrees. Why is this so? And to what end?

In short, life, on the 'phenomenal' level at least, gives the impression of a random succession of 'end products' or an unending process involving an ultimate design, depending on our standpoint.

This impression in fact reflects an over-simpified view of life. For this there are three fundamental reasons.

First, we shall, no doubt, one day realize that the problem is infinitely more complex than we imagine. The purposive thrust in living beings is no less powerful in the 'environment'. We are beginning to appreciate the formidable accomplishments of the primitive forms of life in the atmosphere, in the sea and on the earth's crust, over a period of more than 3,000 million years. For the 20 million or so centuries of the pre-Cambrian, they were actively engaged in mineralization, the creation of proteins and the oxygenation of the earth. In so doing, they paved the way for the biological explosion at the beginning of the Cambrian which would not have been possible without a supporting system. What is more, we could not live without the continued operation of this system. The workers in this planetary chemical industry are micro-organisms. Evolution would also be unthinkable without macroscopic transformations, if the phenomenon which we call the 'environment' were not in some way endowed with a purposive character or used for a specific purpose.

Second, we generally represent evolution as a linear progression engendered by a guiding principle. The fact is that it should be conceived in cyclic terms or, rather, as a system. It is seen as the outcome of a whole series of actions, reactions and forms of regulation on the fourfold plane comprising living particles, organisms, populations and relations between living beings and the environment. In this context, structures and functions are in close correlation; they mutually induce or reinforce each other. There is an old saying that 'the function creates the organ'. Thus, we may see in the retraction of the petal in response to a lack of sunlight an early reflex which was to develop into the nervous system. Are not neural responses a kind of 'pre-vision' from which the eye eventually evolves? This would account for the performance of a single function (vision or flight, for example) by various structures. Admittedly, it does not explain the integration of these responses into the genetic material. For this purpose, however, we should no doubt bear in mind that biological systems are characterized by their openness: by incessant attractions and retractions, propensities and repulsions which start, perhaps, as simple physico-chemical reactions but form the basis for the development of a radically different 'logic' from the 'logic' of matter. In this connection, we feel that certain Neo-Darwinians[3] have over-emphasized the idea of self-preservation and, thus, the internal 'necessity' in biological systems. Surely the latter are equally 'determined' by reproduction? This goes to show that every living being contains a 'principle' which, despite its physico-chemical origins, induces an excentrically oriented 'necessity', by virtue of which it retains perpetual capacity for renewal.[4]

Third, be this as it may, one of the most powerful driving forces in evolution lies, in our view, in the fusion of organisms.[5] We shall develop this point schematically.

When two cells join, two processes begin: (a) the cells no longer have to defend themselves, at least at the junctional cell-wall, against the outside world, so that some of the energy needed for self-defence becomes available for other functions; (b) information flows between them. If we now think in terms of seven cells surrounding an eighth cell, the latter is in an entirely new situation: it no longer has to defend itself against the outside world; furthermore, it gathers the maximum amount of available information which it is able to radiate outwards to the other cells. A process of change begins at this stage (as a matter of fact, biologists have established that, early in the development of embryonic life, differentiation commences in the eighth cell). Another 'model' is that of the virus which enters a bacterium and infuses it with new information. However simplified this schematic approach may be (although it corresponds to reality), it serves to bring out the twofold nature of evolution. 'Necessity'—particularly when considered from a functional point of view—is certainly one of the driving forces of evolution. In this context, however, evolution is seen principally in terms of self-improvement, of moving to a higher level of being. Emergence into qualitatively higher forms of life seems to involve primarily a relationship with the other (by complexification) inasmuch as it is this which makes freedom of necessity.

If this is the case, it is misleading to see Neo-Darwinism and Neo-Lamarckism as opposing views. Life and evolution should be considered in terms of a continuous dialectic between necessity and freedom. By the same token, any application of biological knowledge (if we wish to extend the evolutionary process and pursue the 'logic' of living beings) should both reflect 'necessity' and aspire to preserve, restore and heighten the relational dimension of man.

We again have to ask where this will lead us. Obviously, this question cannot be answered in biological terms. An inkling is possible as to how life acquires a meaning or ceases to have a meaning. We cannot define the meaning of life, since it transcends us by the doors it opens to emergency and relationship to others. In the last analysis, we have to admit that 'life' eludes us; it cannot be explained on a purely physico-chemical basis. The only possible definition of life is grounded in the 'logic' of life:[6] it is a (necessary), 'in-itself', which also ex-sists in the Other (making itself and making the Other, not simply in procreative terms but, in the case of sexualized organisms, in terms of their self-preservation and emergence to a higher level of being). Such a definition means accepting the view that there is an ultimate design to life and at the same time renouncing any attempt to pin that design down, since it too is 'other', extrinsic to life itself.

To our definition of life as an 'in-itself' which ex-sists in the Other, we should add the term 'irretrievably', and this in a number of different senses. First, if only through the action of genetic and other combinations, in life all possible instantiations are tried out, with (as we have already seen) all the hazards of a lottery. If these combinations are incompatible or present a very low degree of compatibility, the result can be a terrible 'mess', extreme suffering, unimaginable wastage. To take the example of Man, at least two-thirds of all engendered organisms do not survive beyond the sixteenth-cell stage of development; and prior to this stage of conception, hundreds of ova and millions of spermatozoa are wasted. All the potentialities of life are or will be exploited, in most cases, in such a way as to result in 'irretrievable' loss. 'Irretrievably' also in that, insofar as it is conducive to the emergence of the Other, the living being actually prepares for its own disappearance. Lastly, by opening itself to the Other, the living being tends to fuse with it and thus lose 'irretrievably' its self-sufficiency; at the same time it paves the way for a new emergence, in which it will itself be sublated.

But if life projects itself irretrievably and the 'logic' of living beings does not provide any clue as to the meaning of life (save in the form of openness to the Other), should we not then seek this meaning in an 'other', another 'order'—the order of reason and freedom? Must we not also acknowledge the futility of the line of thought which we have been developing in an endeavour to derive a 'functional' ethics from the bio-'logical' sphere of reality?

We shall begin by trying to answer the first of these two questions.

The 'life' of the mind

What, then, do we know about human thought? Here, two opposing concepts (or families of concepts) emerge. Some would hold that thought is no more than a manifestation of brain-cell activity. Others maintain that thought cannot be explained purely in physiological terms. The truth of the matter is that the correctness of neither of these views can be ascertained.

Confining ourselves to a number of elementary observations, we know[7] that each nerve-cell, with its myriad synaptic communications, forms a system of unimaginable complexity. And what is to be said about the brain itself, the co-ordinator of these thousands of millions of systems, where the various nerve-centres and even the two hemispheres are mutually regulated? It is doubtful whether man will ever acquire a detailed knowledge of his cerebral activity (apart from the workings of several large 'wholes' and the main forms of regulation). And, even if men were to acquire such a knowledge, they could only be nonplussed at the singularity of each brain and, thereby, find themselves incapable of understanding each other's mental processes (not that this would prevent them from influencing each other's psychological make-up, for better or for worse).

It must be admitted, we find it hard to accept that we do not, or cannot, take cognizance of the substratum of our intelligence. Certain thinkers accordingly jump to the conclusion that thought is no more than a cybernetic phenomenon. Under such a scheme of things, each nerve-cell functions as a computer; its constituent parts can receive signals, classify them and send out other signals in reply, at an amplitude and frequency modulated according to the signals received. Such a 'model' appears to provide a valid account of sensory activity (even though no computer recombines its own 'circuits' as the nerve-cell does with the dendrites). It shows how genetically predetermined programmes of action could be stored or brought into play by appropriate stimuli, and even how sensory impressions could be 'synthesized' (as in a television set) in the form of pictures which are themselves reconstructed or transmitted in response to particular 'needs' (be they purely physiological or behavioural, individual or specific).

Nevertheless, the opponents of what is considered to be a 'mechanistic' or 'materialistic' approach can easily object that the cybernetic 'model' fails to explain the higher manifestations of the expressive, cognitive, simulative or projective faculties. The computer replies, transmits and even translates; however, it cannot create. They echo the well-known saying that the piano does not make the pianist.

We do not feel the two approaches to be incompatible.[8] First, in our view, human thought cannot be considered outside our condition as living beings: there is no such thing as pure reason. In this sense, the 'logic' of the living being is, so to speak, the matrix of thought: that in which and through which thought acquires ex-sistence, even if it aspires to transcend the finitude of this matrix. It is obvious, moreover, that thinking has influenced life, particularly the activity and even the morphology of the brain (which have been progressively refined and complexified).[9] The fact that the converse is also true is evidenced, as we have seen, by the decline in the abstractive powers of a right-handed person in the event of dysfunction in the left hemisphere of the brain. At a deeper level, we see no reason why we should not, for our initial approach, treat intelligence in biological terms. Life is matter; intelligence is life. Originally, it was, perhaps, simply the outcome of internal and external reflexes, retractions and propensities (as in the case of the flower which opens in sunlight), which were registered by our nerve-cells and then systematized and organized (inducing the development and perfecting of our nervous system). It is difficult to draw the line between certain crystallizations and primitive forms of protozoa and, even more so, between such crystallizations and primitive forms of procaryotic cells. In the same way, the first glimmers of intelligence could not have been discerned. However, it is evident that the life of complex organisms incorporates a 'new phenomenon' which institutes another order, a force which differentiates it from matter, and which informs even that which conditions it. By the same token, as the higher manifestations of intelligence attest, the mind transcends the biological order and there is a different kind of force governing its operation.

If we were to pursue our comparison, we would

observe that life involves constant innovation: from a
small number of chemical elements and radicals, there
emerge highly differented cells, polymorphic living
beings. Similarly, from a number of vital functions,
linked to the nervous system, diverse forms of intelligence will emerge and regulate themselves.[10] Could we
perhaps describe matter as the keyboard on which life
plays its words? If this were so, life would have to 'come
from somewhere else' (pre-established in all its potentialities, with the design of man contained in the very first
bacterium). It is not more difficult, if not easier, to
accept the 'creations' of an intelligence emerging from
life (which is not or no longer merely a keyboard, so to
speak) than to accept the 'creations' of life informing
matter.[11]

   It could still be argued that in the above observations
the mind is seen as no more than the sum of our
impressions, whereas reflection pertains to a completely
different domain: it involves stepping outside oneself
and hence a genuine revolution in relation to the reflex.
This we accept. However, we feel we should at this
point consider another revolution: the revolution of
sexualization. As we have seen,[12] sexualization 'throws'
life into the arena of a binary relationship. It shatters
the self-reliance of the individual and destines him, even
morphologically, for the Other. It also leads to the
development of a whole range of signals (smells, sounds,
colours) which are in no way 'necessary' to the self-
preservation of either partner, but become so to enable
both to 'live on' through sexual union and procreation.
At the same time, sexualization constantly induces combinations and recombinations; it opens on to the new
and on to the possibility of emergence. In short, sexualization plucks the individual out of his particularity.

   We could take these points one by one and apply
them to reflective thought. Irrespective of the fact that
reflective thought would seem to presuppose a
reorganization (or new organicity) of the brain (whether
it be in terms of the neo-cortex or the twofold regulation
of the cerebral hemispheres), it is generally accepted
that it develops in mutual relations; that it heightens
the relational nature of the individual (if only through
language); that it involves a constant process of combination and recombination; that it gives rise to the
new (particularly through work); that it transforms

intelligent beings in terms of potentiality and actuality; that it attests that the human condition involves being a particular living creature which has to transcend its own particularity. But sexualization, however revolutionary it was in comparison with binary fission, required life and the agency of living beings in order to come about. If an analogous revolution has come about on the plane of the intelligence, is there any need to seek its origin outside the mind itself? Sexualization is within the 'logic' of life; by the same token, reflection is within the 'logic' of the mind. What, then, is this 'logic'? Our definition of the living being as an 'in-itself which ex-sists in the Other' would seem also to cover the intelligent being. The term 'in-itself' would suggest that the mind to some degree comprehends the special needs of the individual, while the expression 'which ex-sists in the Other' would apply irrespective of whether the Other were a human Other or the noumenal Other of reality, realized or realizable.

The reader should not mistakenly infer that we purport to know what the mind is, any more than we would purport to know what life is. It will never be possible to explain sexualization or the make-up of the brain in purely physico-chemical terms, or even in terms of selection. The phenomenon of life grows more disconcerting with every new advance in the field of biological research. Similarly, our progress in the analysis of the human psyche leads us to conclude that it cannot be understood as a purely physiological phenomenon.

We have treated these questions in some detail with a view to providing a background to our primary concern, which is to consider the proposition that the 'logic' of the mind does not differ essentially from the 'logic' of life. We regard such a proposition as primordial, even though our inability to explain it prevents us from going further than merely affirm or enunciate it. One of the areas of difficulty in accounting for this identity of 'logics' is the interaction between life and the mind (if only through the modification of cerebral activity). But if there is a real identity of 'logics', we have good reason to believe that our knowledge of one dimension can clarify obscurities in the other. We could also reasonably argue that what is 'good' in the one dimension must be equally so in the other.

We are thus able to answer the question which we raised earlier in this chapter: since the phenomenon of life does not indicate the meaning, the purpose, the direction of life, should this not be sought in another order, such as the order of reason (which would invalidate our endeavour to derive a 'functional' ethics from the domain of biology)?

If what we have suggested concerning the 'logic' of the mind is correct, three observations should be made at this point:

Reason can only assign meaning, purpose, direction by entering fully into what it emerges from; thus it must in this case grasp and pursue the 'logic' of life. No rational ethics can abstract from the peculiar exigencies of the evolutionary process. Conversely, the bio-'logical' dimension cannot be pursued in opposition to the peculiar exigencies of reason; above all, it cannot be pursued in opposition to the conditions necessary to the exercise of reason.

While the phenomenon of life does not indicate the meaning of life, except in terms of transition and emergence into the Other, the mind also tends to de-termine things and beings, plucking them out of particularity, to grasp them in their potentiality and actuality. It thereby unveils meaning without ever attaining *the* meaning. In other words, just as life would lay itself open to annihilation if it closed itself to evolution and emergence, so the mind would become meaningless if it did not prepare the way for its own sublation, if it determined itself by obstructing or standardizing the conditions in which it functions (particularly the biological conditions).

Since the living being, like the thinking being, reveals itself as an 'in-itself which ex-sists in the Other', ethics doubtless entails respect for the needs inherent in the preservation and reproduction of life as well as for the inherent needs of reflective thought. In our view, however, the specific object of ethics resides in the transition to ex-sistence and the condition which makes that transition possible: the heightening of the relational dimension. Hence, ethical demands are not to be conceived as arising from without, as it were from some empyrean. Rather, they are to be seen in terms of a 'supplement of logic',[13] modulating the ex-sistence and vice versa.

A parallel conclusion may be reached from a consideration of freedom, the other mode of life of the mind. On this subject we shall confine ourselves to a number of fundamental observations.

A growing number of contemporary thinkers, particularly those influenced by biologism, have come to the point of negating freedom, which is regarded as a mere appearance, although some would go so far as to term it a product of mystification. Thus, Jacques Monod could solemnly maintain that the human mind is 'contained, inscribed in the geometric deformations of several billions of small molecular crystals.'[14] Other thinkers base their theories on the multitude of reflex responses to received impressions. They hold that the number of responses in the human brain is so great that the reactions of the individual to any stimulation are unforeseeable and the illusion of choice or personal creation is thus brought about. But is this not in a sense to confuse the conditions governing the exercise of freedom (and the taking of decisions) and the conscious (or unconscious) movement by which we stand apart from our condition or from a given situation, whether to reject or accept it? Admittedly, there are many different reasons for this standing apart. Freedom cannot be 'isolated'; it cannot show itself in a pure state. By definition, freedom frees itself, and hence presupposes a conditioned state from and within which it emerges. In fact, those who believe in freedom accept that, like reason, it does not demonstrate itself; it shows itself:[15] in relation to it-self and in relation to the Other.

We feel in this regard the presence within us of a secret element, which cannot be explained by biological or social determinisms. We do not, of course, intend to become other than we are, to succeed in being 'more ourselves than we are'; or to attain self-realization outside the 'logic' of the living being or the 'logic' of reason. We do feel, however, that an end cannot be assigned to what we are; that we are in the mode of the future, of an emergence which is at least possible. Of course, such a feeling does not attest to our actual freedom. However, it at least translates a demand which we see as fundamental and constitutive of our being. This demand is inscribed at the very core of life, yet is unobjectifiable (except, perhaps, in certain

extreme situations, such as revolt against degradation
or servitude, or consent to denigrating behests and even
to death). We would repeat that there is no such thing
as pure freedom; it can only appear as a project or
modality of ex-sistence, indicating the presence within
us of another order, which is to the mind what the
latter is to life and life is to matter.

Furthermore, freedom (or our feeling of freedom,
our project of freedom) shows and authenticates itself,
although always unobjectifiably, in consent to the
being of another freedom (as long as we remain in a
relationship involving the possession or subjection of
the Other, our freedom as such cannot be attested).
We recognize, beyond and within the life of the Other,
beyond and in his singularity (after the manner, moreover, of what we have some inkling of within ourselves),
his reality as a subject, by which we mean that in the
Other, as in ourselves, there ex-sists a fundamental
otherness which is both present and in process of
realization. Freedom is consent to the realization of this
otherness.

It appears then that our definition of the living
being and the thinking being is eminently applicable
to the 'Being in the process of giving birth to freedom';
it is an in-itself (consenting to the Self) which ex-sists
in the Other (consents to it). In other words, to act in
freedom ultimately means to apprehend the 'logic' of
life and allow its emergence. Accordingly, those who
deny the presence within us of freedom as a faculty have
good grounds for doing so. But they also have to
recognize that freedom is no more improbable (or
unprovable) than the 'logic' of the living being, and
that this great unknown is to be found at the very heart
of life: it is the very 'logic' of life.

In the light of these various points, we should like
to submit three further observations which may be
useful in establishing the relationship between biology
and ethics.

First, if freedom did not exist, it would have to be
   invented, if only from bio-'logical' considerations.
   And, if freedom does exist, at least as a project, the
   aspiration to freedom is tantamount to an attribu-
   tion of value to that from which it emerges, otherwise it would not exist. Hence, any ethics of freedom
   has to reflect fundamental bio-'logical' necessities.

* The question concerning the subject and the object is extremely important. In philosophy, we do not consider man to be an entity, or an object, while for a very long time the scientific approach has been to seek a defined and simplified objectivity. When in philosophy we say that the human being should be regarded as a subject, this does not mean that we have to give up objectivity; it means that we have to transcend objectivity, that the human being is above and beyond objectivity. Naturally, when he is treated as a mere cipher in a series, man may be regarded as an entity; in this case, however, he is being treated as something inferior to what he actually is. Man must be understood as a subject ... But the person who looks at him is also a subject. Thus, in the case of man, since the person who considers him is also a human being, there are in fact two subjects who confront each other. If there is a real encounter between two human beings, such as between the doctor and his patient, there is *subjectivity* ... Thus, when two human beings confront each other, recognizing the full implications of subjectivity, they transcend the simplified form of being and enter the world of true humanity.

It would seem to me that one way of ensuring a personal human encounter between the field of biology and ethics would be for us to keep alive in ourselves an awareness of this

On the other hand, bio-'logical' science has to foster and respect freedom, since it cannot flourish where there is no freedom.

Doubtless, biology should endeavour to objectify the 'situation' of the living being. But, in so doing, it will only attain the living being as an in-itself; ex-sistence defies objectification. At best we can discern the conditions which make ex-sistence possible; and we can heighten the capacity of the individual to enter into a relationship with the Other, and into his own ex-sistence, according to the dictates of reason. Biology should guard against considering man as a mere object and, *a fortiori*, determining him: man is or tends to be, by virtue of the 'logic' which pervades him, a subject in relationship with other subjects.*

Our conclusion is thus similar to that reached in regard to the question of the 'mind': an ethical approach to the problems posed by biological advance would consist in modulating our understanding of the conditions of biological ex-sistence (and the ex-sistence of the mind) with a project of freedom. In this way, we would endow (or try to endow) the living subject with a 'supplement' of 'logic'.

An approach to the purpose of life

The reader may feel that this chapter has been excessively devoted to questions which are theoretical and abstract (particularly on account of the concise way in which they have been treated), and which are tedious because they are well known. Even the definition which we have returned to time and time again, presenting the living, intelligent and free being as the 'in-itself which ex-sists in the Other' could be a mere truism. However, we feel that our consideration of these questions has enabled us to deepen our insight into the relationship between biology and ethics or, to put it more precisely, to delineate an ethical approach to questions of a bio-'logical' nature.

At this point we would like to summarize the burden of our argument, the full implications of which are now more evident.

First, one proposition has, in our view, emerged as

M

fundamental: the 'why' or 'wherefore' of our biological interventions must be regulated in terms of the 'how' of life, considered in all its complexity. Setting out this aphorism,[16] we stressed the need to exercise caution and view the projects of biological science and the applications of biological progress critically, in the light of their compatibility with the inherent needs of biological 'systems' (in terms of symbiosis and synchrony). But the ultimate reason for regulating the 'why' in terms of the 'how' resides, in our opinion, in the fact that at whatever level the phenomenon of life is envisaged (physiology, mind, freedom), its only discernable purpose is to ex-sist, to facilitate the advent of the 'Other'. Accordingly, it only endures and acquires meaning insofar as the 'Other' emerges actually or potentially on to a higher level of being. However, it does not pertain to us to predetermine this otherness, this Being of the Other.

Second, it follows that we should aim to de-termine life. This proposition, which forms a corollary to the preceding proposition, is to be understood in relation to the twofold dynamic process which, in our view, underlies evolutionary progress. It is our submission that evolution advances in the direction of improvement (existence better and more fully attuned to the 'in-itself' of the living being, enhancement of the internal needs and controls) and also in the direction of higher levels of being (with the emergence of qualitatively higher forms of life). Admittedly biology must seek improvement, must strive to combat individual disorders and provide for the essential needs of mankind at large, in line with the modern tendency to seek perfection in the state in which we are. In this connection, we should remember that in the field of biology the term 'perfection' means sclerosis; at one end of the scale, the 'perfect' example of life is the ant-hill (at least as commonly depicted), where all the members are strictly conditioned and specialized on a purely functional basis. Biology would become ludicrous if it aspired to involution instead of evolution and emergence. We would reiterate that evolution and emergence cannot be predetermined: at the very most, we can strive to contrive or foster the conditions which, in our view, have made such emergence possible and are indispensable to any further emergence.

subjectivity, for the world is more than a mere collection of entities: it is, rather, a horizon—which also encompasses subjectivity . . . Hence, we should strive to develop a new methodology, which will bring together and combine subjectivity and the objectivity of science . . .—*Extract from the statement made by Professor Jamalpur in the general discussion.*

Third, the basic precondition in this context is to preserve, restore and, if possible, heighten the relational dimension of living beings insofar as union, compenetration and organization enable living beings to transcend their own particular needs. We would refer here to our various observations concerning sexualization, procreation and the urgent need for social integration. This proposition in fact involves the essence of man as such, insofar as the latter can only attain self-realization in terms of his own attributes (his intelligence, his project of freedom) by opening himself to the Other. It undoubtedly also involves the very essence of life which (it is a fact) evolves and advances in mutual relationships.

Fourth, it is clear, however, that this openness to the Other, even when it implies man's incorporation in society and, accordingly, the acceptance of certain constraints and controls, entails respect for the singularity of individuals. We have already expressed our view that standardization would lead to an impoverishment of mankind, and impair man's evolutionary potential, while also endangering his adaptability. We would add that the project of freedom (consent to self, consent to the Other) may be regarded as an intrinsically biological phenomenon insofar as it embodies the 'logic' of living beings.

Lastly, it is essential to accept the full implications of the historicality of life, even if this historicality presupposes an integration of death. There is surely no need to repeat that life is reproduction or, to put it better, procreation; that it 'is passed on' from one living being to the other; that it is evolutive and capable of emergence; that, if it were to cease at the present stage of individual existence at a certain form of development of the potentialities of the individual, at a certain period in the history of a society, it would become a meaningless phenomenon. At first sight, it no doubt appears to be a meaningless phenomenon, since evolution advances blindly and involves many failures and disorders. This much is evident. But it is no less evident (and becomes obvious in relation to the mind and the project of freedom) that the logic which pervades life 'authorizes' a continuing process of sublation. Biology purports to struggle against everything in life which represents a lack (a breach or a deficiency) of logic, and to strengthen the living 'in-itself'. If it endeavours to attain this object by mortgaging the future of individual living beings, coming generations or mankind at large, bio-'logical' science will become quite illogical.

Thus, biology can be seen as the science of the relational, the science of emergence, the science of history. But, viewed in this perspective, it obviously may, and indeed should, be challenged by related scientific fields such as psychology, ethology and sociology. It would also be a positive step if biology were to take a critical view of advances in these fields. A complex interdisciplinary or transdisciplinary approach to man as an 'organic totality' appears to be more and more necessary, as Professor Anguelov stressed at Varna.[17] Ultimately, the phenomenon which we call 'life' is incomprehensible; it is actually and potentially beyond our mastery, and cannot be reduced to an aggregate of physiological data. The mind and freedom also embody the 'logic' of life.

If life is beyond our mastery, how are we to discern the 'good'? The reader has surely perceived the conclusion around which we have been centring our argument. In our view, ethics has to deal with the various problems posed by biology by taking a critical look at biological research and practices in relation to the logic which pervades the living being. We must respect and strengthen the living being as it is or should be in-itself, but in seeking improved conditions of existence we must also aspire to higher levels of being. In the last analysis, the specific object of ethics is undoubtedly to project life into ex-sistence. Such a conclusion is bound to appear laughable since it has a perhaps deceptively familiar ring about it. The reader may wonder why so many pages have been needed to reach such a conclusion. In particular, it does not see biology as having any determined or determinable end, purpose or direction.

As we have said, there is no science of ex-sistence; at the very most it is possible to pinpoint a number of conditions for ex-sistence to come about. Historically, ex-sistence in the Other entails commitment to a definite position, a 'logic' to be grasped in its individual and its social implications; but it is also an adventure. Admittedly, the adventurous aspect of ex-sistence is repugnant to us and causes us anxiety (perhaps we are mysteriously put on our guard by the vagaries of life). We are tempted, then, to appeal to an 'outside' science to predetermine and normalize this ex-sistence (and, accordingly, biology). Certain thinkers would like to see ethics perform such a function. By the same token, ethics is also expected to grasp the Absolute and promulgate principles which are universally and timelessly valid. This is to forget that, if such principles could be established, they would only be perceptible through a human and, therefore, historically situated formulation. These principles could only apply to pure freedom, whereas we are freedom incarnate. No more could we appropriate them absolutely than we can appropriate our own project of freedom. Whether we like it or not, it falls to each of us to constitute himself as a subject of and through the adventure of ex-sistence.

Where will this adventure of ex-sistence take us? To seek to qualify oneself to ex-sist in the Other commits us irretrievably to a limitless quest for sublation. At least it would appear so. In reality, however, anyone who appropriates this logic gains access to Meaning. For what can reason tell us of the Absolute or, to put it better, of God, if we have faith in Him, other than that He is (in the Judaeo-Christian tradition) the Living God—the in-itself which ex-sists in the Other, giving ex-sistence to every being. It falls to us to appropriate this Logos: the Future, the End, the Beyond transcend us.[18]

Recommendations

It is possible that, throughout this study, our argument may have been clumsily conducted. It may also be open to criticism on the grounds of various errors or omissions; overemphasis of the relational, emergent, historical character of life; or excessive extrapolation from the compenetration of living beings. Some may feel that more attention should have been given to other 'values'. Furthermore, various problems raised by recent advances in biology have been passed over in silence while most of those referred to have been no more than touched on, with a view to illustrating a point. Lastly, we shall undoubtedly be taken to task for paying inadequate heed to the dissimilarities between the socio-cultural and economic situations of the highly medicalized industrial countries and developing countries. So be it. We warned the reader at the outset of this work that it is merely an essay. Our main purpose has been to encourage further discussion and suggest possible themes for other conferences and seminars.

In point of fact, the Varna meeting—and particularly the fact that the moralists present there remained on the *qui vive*—brought out the urgent need to define the scope of ethical science. Has it an independent aim and method? Should it take the form of a synthesis confronting, arbitrating and regulating in relation to each other the various 'logics' deriving from biology, sociology and economics? In relation to what, however, would such arbitration and regulation function? Ethics has long been associated, implicitly or explicitly, with various philosophical or religious systems which are today questioned or simply considered obsolete, at least in their traditional form. The question is, then, whether—and, if so, on what basis and by what means— mankind will at any rate credit man with a certain intentionality or overall design. This is a crucial question if we wish to avoid mortgaging our future in pursuit of a mere improvement in our conditions of existence. The most serious problems are perhaps not so much those posed by advances in the field of biology as those resulting from the lack of moral and philosophical thinking about the implications of scientific progress. The Final Report of the Varna meeting called attention to this shortcoming:

'The diversity and the complexity of the problems touched upon in the debates have shown:

'1. The effort that philosophy as a whole has to make to take into due account the new vision of the world resulting from the general progress of science;

'2. More precisely, the urgency of developing a new ethic in order to put the capabilities of science, particularly those of molecular biology, at the service of human rights and the general good of societies in both industrially developed and developing countries;

'3. But also the fact that scientists must endeavour, in accordance with the desire they themselves have expressed, to enter into discussion with moralists in order to think together and to help each other clarify their ideas and adopt standpoints which take into account both the remarkable capabilities of science and the moral requirements founded on human rights. No doubt it is because these three preparatory steps have not yet been taken that two important questions referred to in the working document have not been answered conclusively (and in the current state of scientific and philosophical thinking answers are perhaps not easily found, although the debates show that progress is already being made in this direction). These two questions are as follows:

'(a) first, the relationship to be established today between scientific and ethical research the conclusions of which have, up to now, merely followed each other consecutively without ever integrating, with morals assessing post factum the applications of scientific discovery, whereas logic and common sense alike would suggest that their conclusions and findings should be arrived at simultaneously, or even by a single, combined effort;

'(b) secondly, the problem of the definitions which should stem from this relationship with a view to formulating the main points of an ethics for man and society, in both industrially developed and developing countries.'

The aims thus set out in the Final Report will only be achieved if a considerable amount of work is rapidly undertaken. In the first place, interdisciplinary studies will have to be carried out. Then educators will have to take the results of this concerted approach to the problems posed by scientific advance into account so as, at least, to introduce a new approach to ethical questions (including questions of professional ethics) in medical schools and biological research training institutes.[19] Lastly, popularization programmes will have to be organized[20] to give the general public a keener perception of the social responsibilities of the individual, with regard to his own life, his offspring's life and the life of mankind at large. Many of those attending the Varna meeting stressed the urgent need for this threefold work.[21] Furthermore, eight of the ten recommendations addressed to the Director-General of Unesco convey the wish that this work should be encouraged and furthered.[22] We feel it appropriate to

quote these recommendations in the concluding
section of this study:

'The meeting of experts suggests that the Director-
General of Unesco:

'(a) continue and strengthen the programme on science
and ethics through keeping the ever-changing
subject of ethics in relation to the sciences under
constant review by sponsoring further international
and inter-disciplinary working groups in developed
and developing countries and, through publications,
disseminate world wide the authoritative recommendations of these groups. Such encouragement
of good and discouragement of bad practices and
procedures would contribute to the improvement
of the quality of life;

'(b) help define the concept of a new functional ethics
and encourage research in ethics;

'(c) establish in close collaboration with WHO and the
CIOMS a standing committee of scientists and
philosophers to monitor the applications of
biological discoveries;

'(d) include the teaching of a modern functional ethics
related to scientific discovery in Unesco's educational and scientific programmes and curricula;

'(e) encourage and keep under continuous observation
the provision of courses of ethics for all university
and technical college students;

'(f) allay public anxiety and raise awareness of biological discovery and its ethical implications
through the mass media inside and outside Unesco
through accurate, up-to-date and honest information;

'(g) include the study of biological problems, and in
particular those relating to ecology, from both the
scientific and ethical points of view among the
main goals of Unesco's programme on the human
implications of scientific advance;

'(h) encourage biologists, social scientists and philosophers to collaborate with a view to improving
the quality of life for peoples in both developing
and developed countries; . . .'

For our part, as we indicated in our Foreword, in
presenting these (what to some readers may seem
unduly personal) reflections on the Varna meeting we
have borne the second of these recommendations

fixedly in mind. If we have succeeded in helping to 'define the concept of a new functional ethics', we shall feel that we have achieved our fundamental aim.

Notes

1. cf. above, pages 18 and 30 et seq.
2. These basic questions were not debated at Varna. The arguments set out here do not, therefore, necessarily reflect the opinions of any of the participants.
3. We are thinking here of the work, for example of Jacques Monod. However, other scientists of the same 'school' (François Jacob, for instance) have emphasized the importance of the 'openness' of the living being.
4. In this context, the work of Motoo Kimura may be regarded as of considerable importance.
5. We have developed this theme in *Cherchant qui Adorer*, op. cit., Part I, Chapter 1.
6. We are here applying the bacteriological theories of F. Jacob to all living beings: cf. above, p. 114.
7. cf. above, pages 96–98.
8. These ideas are dealt with at greater length in *Cherchant qui Adorer*, op. cit., Part I, Chapter 2.
9. In this connection, volumes could be written about certain types of education which either devote excessive attention to a specific faculty or simply neglect it and leave it dormant.
10. We believe in animal intelligence.
11. For that matter, what intelligence would be capable of 'creating' a brain?
12. cf. above, pages 115–118.
13. This expression is inspired by the *supplement d'âme* ('supplementary infusion of soul') which Bergson wished to see brought into a world dominated by technical progress (cf. *Les Deux Sources de la Morale et de la Religion*, Chapter IV, p. 1239, Paris, Presses Universitaires de France.).
14. Inaugural lecture at the Collège de France.
15. Pascal had already noted this in his *Pensées* (No. 268, ed. Brunschvig) in reference to reason.
16. cf. above, pages 38–39 and 54–55.
17. cf. above, page 30.
18. These suggestions form the central theme of our work *Cherchant qui Adorer*, op. cit., to which we would refer readers seeking a more detailed treatment of these points.
19. Professor Mercier observed that ethics (including professional ethics) have practically disappeared from the curricula of medical schools in German-speaking and English-speaking countries, that in France the nature of teaching in this subject varies greatly from institution to institution and that, although the subject is still compulsory in Italy, the courses given in it are out of date, in both form and content.
20. Dr Koen suggested to the participants at the Varna meeting that they 'actively encourage the establishment of institutions or bodies designed to provide a responsible, matter-of-fact information service to the public on advances in biology and the personal responsibilities and requirements stemming from them'.
21. The following participants in particular spoke on this question:

Dr Akande, Professor Jamalpur, Dr Kayper-Mensah, Dr Mroueh and Dr Wolstenholme. Professor Kholodilin had already emphasized this point in his introductory speech.

22. The participants also suggested that the Director-General of Unesco: 'in co-operation with WHO and FAO, encourage biologists to concentrate on means to solve the two main problems of mankind: health and hunger; and follow closely current and future research in molecular genetics, keeping in mind the dangers implicit in certain new techniques in this field of research'.

Appendix

List of participants

Dr E. O. Akande, Dean, Faculty of Medicine, University of Ibadan (Nigeria).

Dr Günter Altner, Professor of Theology and Biology, Forschungsstätte der evangelischen Studienstiftung, Heidelberg (Federal Republic of Germany).

Professor Stéfan Anguelov, Directeur, Centre Unifié de Recherche et de Préparation de Cadres en Philosophie et en Sociologie auprès de l'Académie Bulgare des Sciences, Sofia (Bulgaria).

Professor V. V. Bezrukov, Researcher, Physiological Laboratory, Institute of Gerontology, Academy of Medical Sciences, Kiev (Ukrainian Soviet Socialist Republic).

Professor Kiril Bratanov, Directeur, Institut de Biologie et Pathologie de la Reproduction Animale, Académie Bulgare des Sciences, Sofia (Bulgaria).

Professor L. K. Dramaliev, Professeur d'éthique, Université de Sofia (Bulgaria).

Professor D. Waclaw Gajewski, Instytut Botaniki, Wydzial Biologii Uniwerstytetu Warszawskiego, Warsaw (Poland).

Professor Bahram Jamalpur, Philosophy Department, School of Letters, University of Tehran (Iran).

Dr Vasken Der Kalustian, Associate Professor, Department of Paediatrics, American University Hospital, Beirut (Lebanon).

Dr A. W. Kayper-Mensah, Director of Culture, Ministry of Education, Youth and Culture, Accra (Ghana).

Dr L. Koen, Director, Stichting Bio-Wetenschappen en Maatschapij, Leiden (Netherlands).

Professor P. Malek, Institute of Clinical and Experimental Medicine, Prague (Czechoslovakia).

Dr Adnan Mroueh, Associate Professor, Department of Obstetrics and Gynaecology, American University Hospital, Beirut (Lebanon).

Dr Shankar Narayan, Acting Secretary, University Grants Commission, New Delhi (India).

Professor Bruno Ribes, Editor of the review *Études*, Paris (France).

Dr Harmon Smith, Professor of Moral Theology and Community Health Science, Duke University, Durham, North Carolina (United States).

Professor M. E. Vartanian, Director, Pathophysiological Laboratory, Institute of Psychiatry, Academy of Medical Sciences, Moscow (U.S.S.R.).

Dr Louis Verhoestraete, Executive Secretary, Council for International Organizations of Medical Sciences (Belgium).
Dr G. E. W. Wolstenholme, Director, Ciba Foundation, London (United Kingdom).

Invited papers

Dr J. Bril, Société de Vente d'Aluminium, Centre de Recherche Pechiney, Voreppe (France).
Professor V. L. Deglin, Research Director, I.M. Sechenov Institute of Evolutive Physiology and Biology, Academy of Sciences, Leningrad (U.S.S.R.).
Professor N. P. Dubinin, Director, Institute of Genetics, Academy of Sciences, Moscow (U.S.S.R.).

Observers

United Nations: Mr Peter Stone, Editor-in-Chief, Newspaper Development Forum, United Nations, Geneva (Switzerland).
World Health Organization and Council for International Organizations of Medical Sciences: Dr L. J. Verhoestraete, Executive Secretary, Council for International Organization of Medical Sciences, Geneva (Switzerland).
International Council for Philosophy and Humanistic Studies: Professor M. A. Mercier, Secretary-General, International Federation of Societies for Philosophy, Berne (Switzerland).
International Union of Scientific Workers: Professor Kiril Bratanov, President, Bulgarian Union of Scientific Workers, Varna Section (Bulgaria).
Radio Hilversum: Mr M. Visser, Biologist, Philosophical Institute, University of Leiden (Netherlands).
Mr Brandt, The Netherlands Broadcasting Foundation, Hilversum (Netherlands).

Unesco Secretariat

Dr Kiril Delev, Division of Philosophy.
Professor Alexander Kholodilin, Division of Scientific and Technological Development.
Dr Wolfgang Schwendler, Division of Philosophy.